Radiographic Anatomy
A Working Atlas

Radiographic Anatomy

A Working Atlas

Harry W. Fischer, M.D.

Former Chief of Radiology
Strong Memorial Hospital;
Professor of Radiology and
Former Chairman, Department of Radiology
University of Rochester School of Medicine and Dentistry,
Rochester, New York

McGraw-Hill Book Company

New York   St. Louis   San Francisco   Colorado Springs   Oklahoma City
Auckland   Bogotá   Caracas   Hamburg   Lisbon   London   Madrid   Mexico
Milan   Montreal   New Delhi   Panama   Paris   San Juan   São Paulo
Singapore   Sydney   Tokyo   Toronto

RADIOGRAPHIC ANATOMY: A Working Atlas

1234567890 HALHAL 89321098

ISBN 0-07-021089-6

This book was set in Optima by York Graphic Services, Inc.; the editors were
William Day and Peter McCurdy; the production supervisor was
Robert R. Laffler; the book was designed by José Fonfrias; the cover was
designed by Joe Cupani.
Arcata Graphics/Halliday was printer and binder.

LIBRARY OF CONGRESS CATALOGING-IN-PUBLICATION DATA

Fischer, Harry W.
Radiographic anatomy: a working atlas/Harry W. Fischer.
p. cm.
ISBN 0-07-021089-6
1. Anatomy, Human—Atlases.   2. Radiography, Medical—Atlases.
I. Title.
[DNLM: 1. Anatomy—atlases.   2. Radiography—atlases. WN 17F529r]
QM25.F48 1988
611'.0022'2—dc19
DNLM/DLC
for Library of Congress
88-9253
CIP

To Jason, Nathaniel, Jessica,
Cooper, Harry William IV,
Anna, and Catherine.
I would be pleased if one or more of you
would find this book useful.

Contents

Preface

For some time I have thought that a book such as this would be appropriate for medical students in this modern age of medical imaging. Beyond all doubt, the student who intends to be a physician must first learn intimately the structure of his or her patient. The study of human anatomy is quite logically placed first in the medical school curriculum; it is the base for all the other knowledge the future physician must acquire. Upon knowledge of human anatomy is built knowledge of physiology, microbiology, biochemistry and biophysics, pharmacology, and pathology, followed by all of the clinical curriculum. The student must study and learn the anatomy of the human body by cadaver demonstrations and dissections supplemented by all kinds of teaching aids, such as models, books, charts, and slides. However, I have seen a great need for students to become acquainted from the outset with human anatomy as they will most often encounter it in the remainder of their medical careers—that is, in the form of the images of radiology, ultrasonics, nuclear medicine, and magnetic resonance. Although every physician will see the external appearance of patients, all physicians, following completion of the gross anatomy course, will be highly dependent on the anatomic information provided by the images derived from the techniques of modern medical imaging, which employ one form of electromagnetic radiation or another. Specialists will see internal anatomy to some degree—the endoscopist will see the lumen of the respiratory or alimentary tract, and, during specific operations, the chest surgeon will see the contents of the thorax, the abdominal surgeon the abdominal contents, and orthopedists the muscles, joints, and fascial planes—but they too will depend maximally on the images of radiology and associated techniques for initial diagnosis and understanding of the patient's illness, injury, or congenital anomaly and for assessing the progression of the condition and the worth of the treatment. This book, therefore, aims to provide the student with knowledge of human anatomy as he or she will see it in the practice of medicine.

An early stimulus to preparation of this book came from seeing the *Elementaer Rontgenanatomi* by my friend and colleague, the late Dr. Hans Henrik Jacobsen of Copenhagen, first published by Munksgaard in 1973. However, this small paperback book and some of the more recent and larger hardcover competitors I find either limited to x-ray imaging or insufficiently involved with computed tomography, nuclear medicine, magnetic resonance, and ultrasonic imaging.

I wished in my book to stress x-ray imaging (including the various contrast examinations) and computed tomography, with lesser emphasis on magnetic resonance and nuclear imaging. I excluded ultrasonic imaging because I wanted to employ images which best illustrate gross anatomy and do not confuse the student. At the same time, my selection favors the more frequently used clinical images rather than the rarely used ones.

For the computed tomography images, I chose cross sections to show the vital organs and structures without employing every cross section at 0.5- or 1.0-cm intervals, as is the case in clinical practice. Since I felt that normal anatomy was more accurately demonstrated by CT than by magnetic resonance imaging, MR images were used mainly for imaging planes not currently available with CT, such as the sagittal and frontal planes.

The omission of images of the intracranial and intraspinal structures was by design. Neuroanatomy is taught as a separate discipline and course in most medical schools, and thus I have made no attempt to include it here. Perhaps it requires a separate, related imaging book.

I wanted to do without accompanying explanatory text, letting the images speak for themselves, but the student may find one of the standard anatomy texts helpful in the learning process.

Lastly, I wanted to keep the book affordable for the student. Initially the book was designed mainly for first year medical students. I see it as useful to all physicians who need to know anatomy as seen in modern medical images.

ACKNOWLEDGMENTS

Alyce Norder and John J. Allen of Lakeshore Graphic Arts made all the drawings of the anatomic structures. I am indebted to them for their excellent work and cooperation.

For assisting me to obtain the normal images shown in the book, I'd like to express my thanks to the following members of the faculty of the Department of Radiology of the University of Rochester: Dr. Oscar Gutierrez for the angiograms, Dr. James Lovelock for the arthrograms, Dr. John Wandtke and Dr. Donald Plewes for chest and CT chest images, Dr. Joyce Janus for the mammography images, Dr. David Weber for the nuclear scans, Dr. Jack Colgan and Dr. Jovitas Skucas for the gastrointestinal images, Dr. Robert Spataro for the urethrograms and hysterosalpingograms, Dr. James Manzione for the CT images of the neck, Dr. John Thornbury and Dr. Stan Weiss for the computed tomography and Dr. Thornbury for the magnetic resonance images. Dr. Martin Kleinman of the Department of Medicine kindly supplied the ERCP image. From the Strong Memorial Hospital Department of Radiology technology staff, Beverly Skutt and Dan Prosser were very helpful in finding the images for the book.

Most of the images came from the teaching files of the Department of Radiology of the University of Rochester School of Medicine and Dentistry. The files of the Eastman Kodak Company supplied some of the images, and I am grateful for these and for the assistance of Mr. John Cullinan.

Introduction

In radiography of the living man, identification of the images is based on the path taken by the central x-ray as it passes through the body part. A *posterior-anterior (PA) view of the chest* is one in which the x-ray enters from the back, or posterior, surface of the chest, passes through the chest, exits from the front, or anterior, surface, and then reaches the x-ray film where it interacts with the photosensitive emulsion. The emulsion is then developed chemically to bring out the image. In a PA view, the central ray coming from the x-ray tube is perpendicular to the film and is centered to the center of the part being examined (Fig. 1). An *anterior-posterior (AP)*

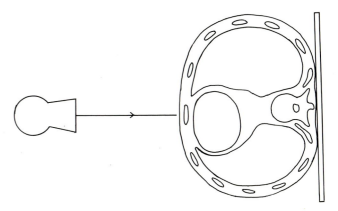

Fig. 2 AP view of the chest.

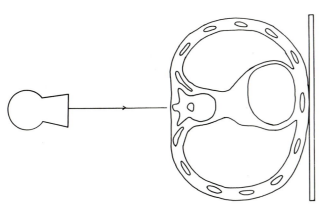

Fig. 3 Lateral view of the chest.

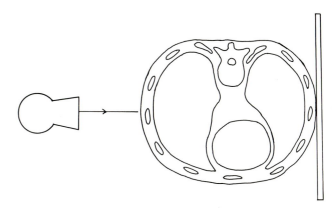

Fig. 1 PA view of the chest.

view is one in which the central x-ray enters from the anterior surface of the trunk or body part, exits from the posterior surface, and then reaches the film (Fig. 2). A *lateral view* is one in which the central x-ray enters one side of the trunk or body part and exits from the opposite side (Fig. 3). These three projections, PA, AP, and lateral, suffice for the majority of images in clinical practice. *Oblique views* are those in which the central x-ray passes obliquely through the trunk or body part, at an angle approximately halfway between a PA and a lateral view. By convention, a *left anterior oblique view of the*

chest is one in which the x-ray obliquely enters the right back of the subject and exits from the left front (Fig. 4).

For radiology of the extremities, AP, lateral, and oblique are the usual views. For the knee and heel, the *tangential view* is also used. For the knee, the tangential view is termed the *sunrise*, or *sunset, view*. An *axillary view* of the shoulder has the central ray directed medially upward into the axilla with the x-ray film above the shoulder. The *navicular view* of the wrist is a special

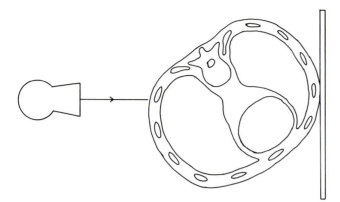

Fig. 4 Left anterior oblique view of the chest.

view based on positioning the hand in strong ulnar deviation to better visualize the navicular bone.

Directing the central ray through the *open mouth* in the AP projection is used for better visualization of the atlas, axis, and the odontoid process of the cervical spine.

For *the skull,* a number of different views employ different orientations of the skull in relation to the central ray and the x-ray film. The *PA view* of the skull or sinuses has the central ray angled 15° to the orbital-meatal line (Fig. 5). The *lateral view* of the skull has the patient's head turned to the side against the film, while the central ray is directed perpendicular to the film (Fig. 6). The

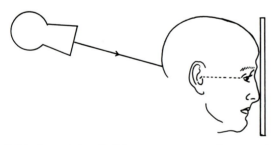

Fig. 5 PA view of the skull [→—: represents the central ray; – – – – – – –: represents the orbital-meatal (OM) line, from the angle of the eye to the external meatus of the ear].

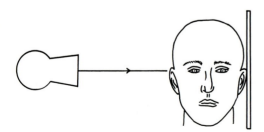

Fig. 6 Lateral view of the skull.

basal (submentovertical) view of the skull has the central ray directed perpendicular to the film and the head tilted back so that the orbital-meatal line is parallel to the film (Fig. 7). The *occipital,* or *Towne, view* of the skull and

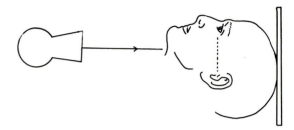

Fig. 7 Basal view of the skull (see Fig. 5 for key).

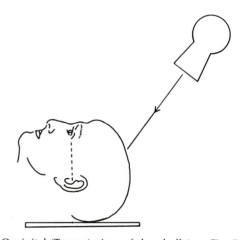

Fig. 8 Occipital (Towne) view of the skull (see Fig. 5 for key.)

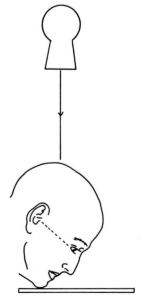

Fig. 9 PA (Waters) view of the maxillary sinus and facial bones (see Fig. 5 for key).

Fig. 11 Law view of the mastoid.

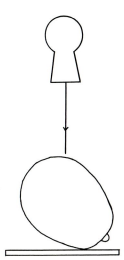

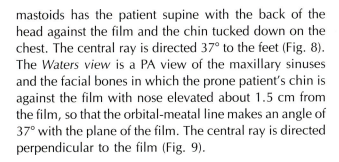

Fig. 10 Stenvers view of the mastoid.

mastoids has the patient supine with the back of the head against the film and the chin tucked down on the chest. The central ray is directed 37° to the feet (Fig. 8). The *Waters view* is a PA view of the maxillary sinuses and the facial bones in which the prone patient's chin is against the film with nose elevated about 1.5 cm from the film, so that the orbital-meatal line makes an angle of 37° with the plane of the film. The central ray is directed perpendicular to the film (Fig. 9).

Finally, there are a number of special views of *the mastoids*. The *Stenvers view* of the mastoids has the central ray angled 12° to the head of the patient while the prone patient's head is turned 45° to the film (Fig. 10). The *Law view* of the mastoids and temporomandibular joints has the central ray angled 15° to the patient's feet and 10° to 15° to the face, while the prone patient has the side of his or her head against the film (Fig. 11).

Radiographic Anatomy
A Working Atlas

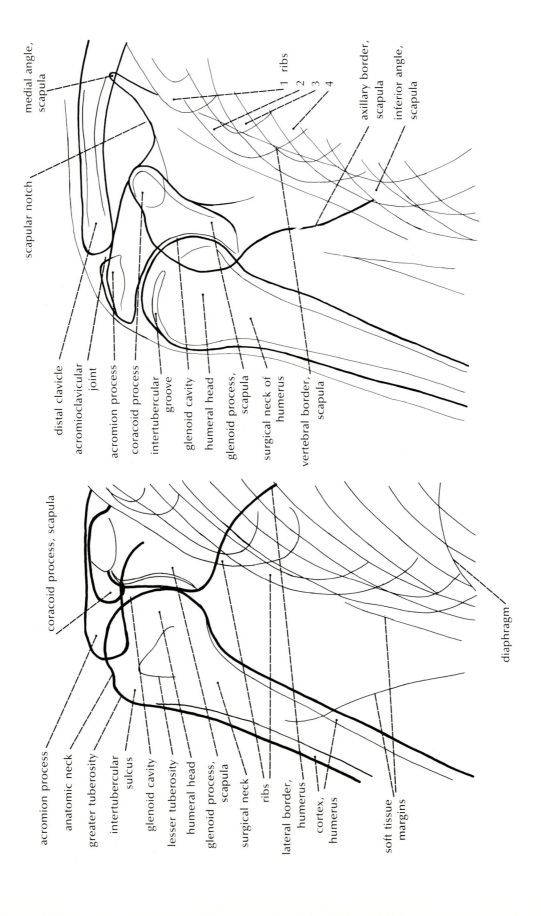

medial angle, scapula

ribs 1 2 3 4

axillary border, scapula

inferior angle, scapula

scapular notch

distal clavicle

acromioclavicular joint

acromion process

coracoid process

intertubercular groove

glenoid cavity

humeral head

glenoid process, scapula

surgical neck of humerus

vertebral border, scapula

coracoid process, scapula

diaphragm

acromion process

anatomic neck

greater tuberosity

intertubercular sulcus

glenoid cavity

lesser tuberosity

humeral head

glenoid process, scapula

surgical neck

ribs

lateral border, humerus

cortex, humerus

soft tissue margins

1/Shoulder (humerus)

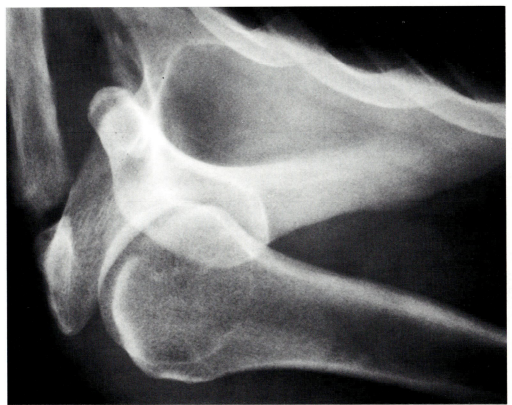

1-2 Shoulder, AP view, humerus in internal rotation.

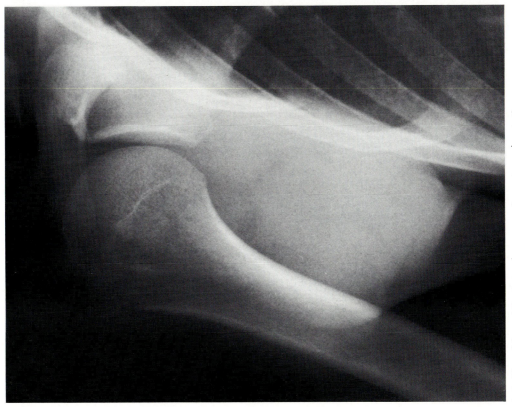

1-1 Shoulder, AP view, humerus in external rotation.

1/Shoulder; Scapula

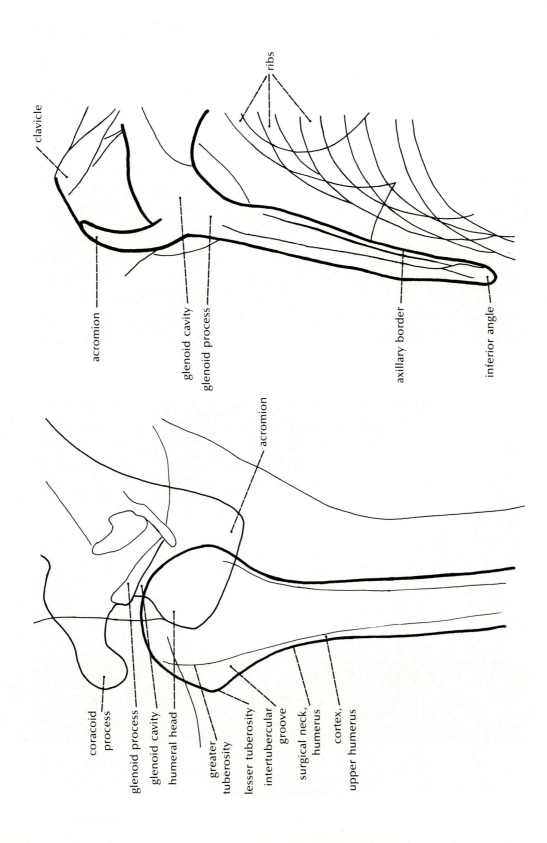

clavicle

ribs

acromion

glenoid cavity

glenoid process

axillary border

inferior angle

acromion

coracoid process

glenoid process

glenoid cavity

humeral head

greater tuberosity

lesser tuberosity

intertubercular groove

surgical neck, humerus

cortex, upper humerus

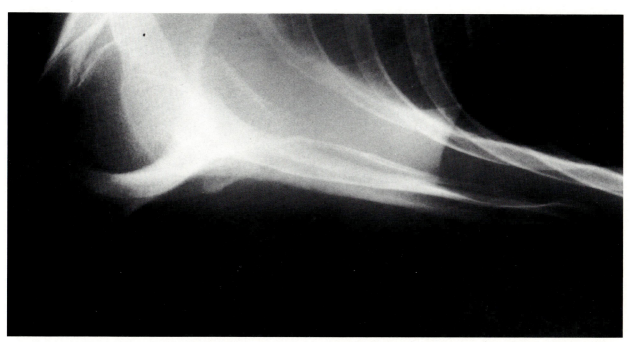

1-4 Scapula, tangential view.

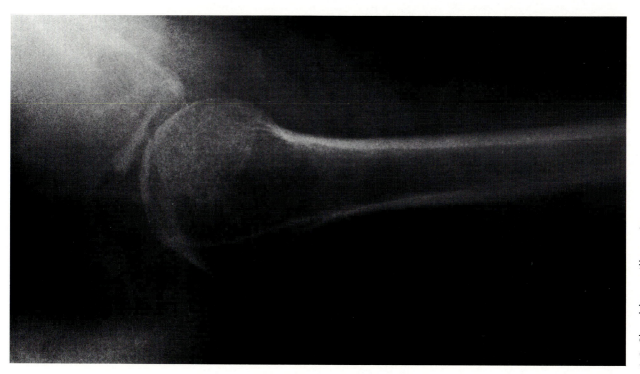

1-3 Shoulder, axillary view.

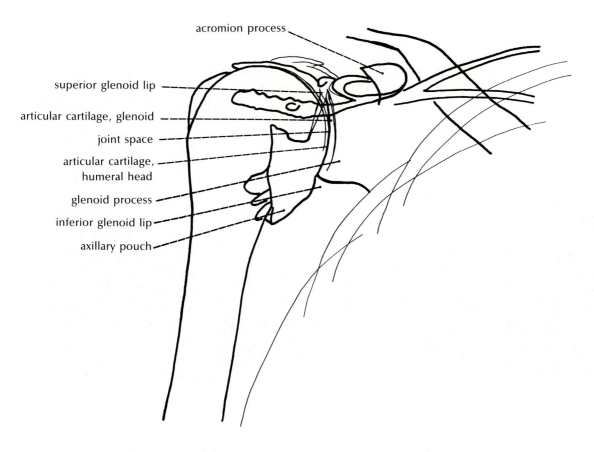

acromion process

superior glenoid lip

articular cartilage, glenoid

joint space

articular cartilage,
humeral head

glenoid process

inferior glenoid lip

axillary pouch

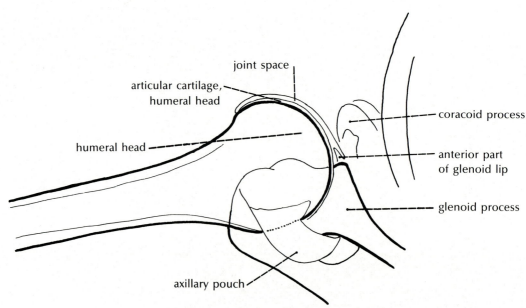

joint space

articular cartilage,
humeral head

humeral head

coracoid process

anterior part
of glenoid lip

glenoid process

axillary pouch

1/Shoulder, Arthrography

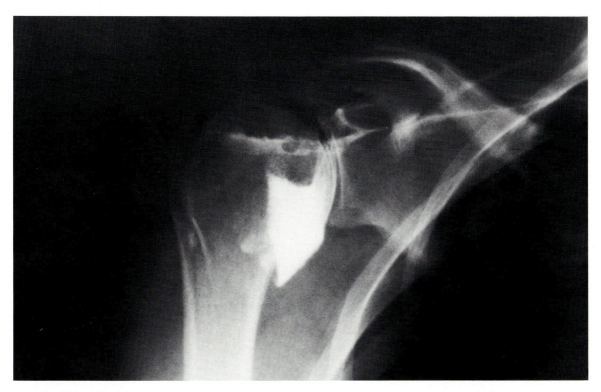

1-5 Shoulder, double-contrast arthrogram, AP view.

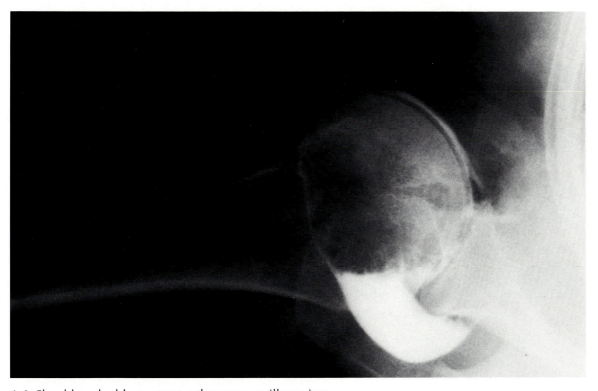

1-6 Shoulder, double-contrast arthrogram, axillary view.

1/Upper Arm (humerus)

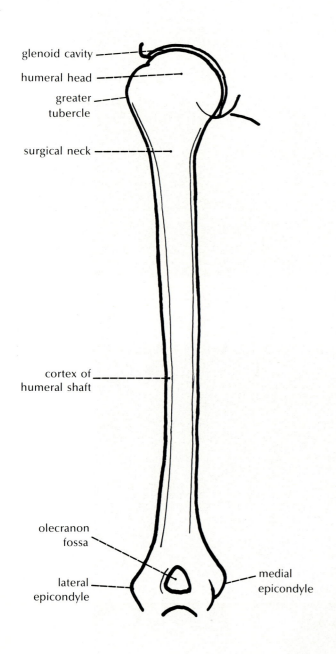

glenoid cavity

humeral head

greater
tubercle

surgical neck

cortex of
humeral shaft

olecranon
fossa

lateral
epicondyle

medial
epicondyle

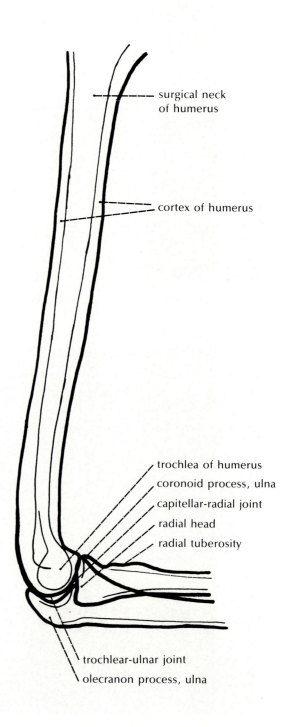

surgical neck
of humerus

cortex of humerus

trochlea of humerus

coronoid process, ulna

capitellar-radial joint

radial head

radial tuberosity

trochlear-ulnar joint

olecranon process, ulna

1/Upper Arm (humerus)

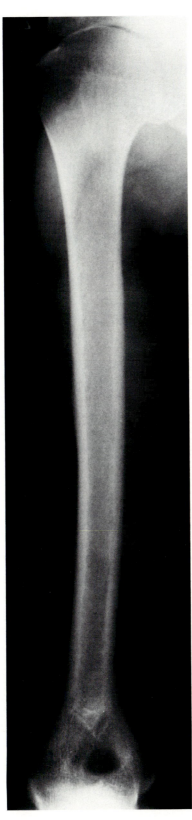

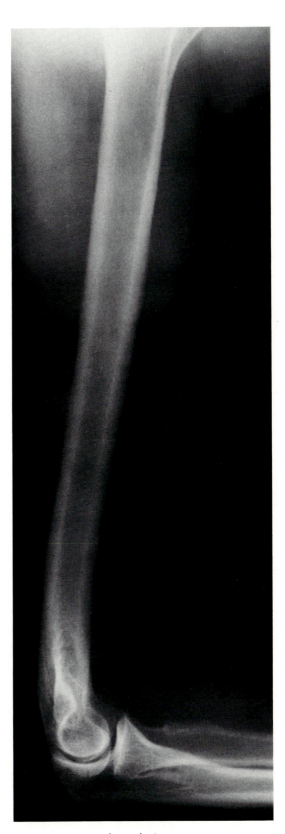

1-7 Upper arm, AP view.

1-8 Upper arm, lateral view.

1/Elbow; Finger

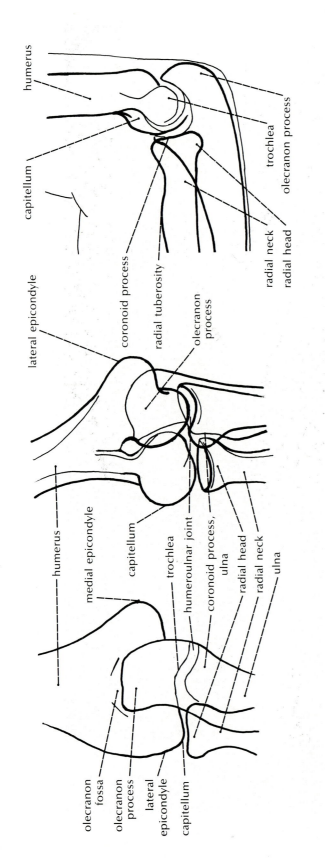

humerus

capitellum

trochlea

olecranon process

radial neck

radial head

lateral epicondyle

coronoid process

radial tuberosity

olecranon process

humerus

medial epicondyle

capitellum

trochlea

humeroulnar joint

coronoid process, ulna

radial head

radial neck

ulna

olecranon fossa

olecranon process

lateral epicondyle

capitellum

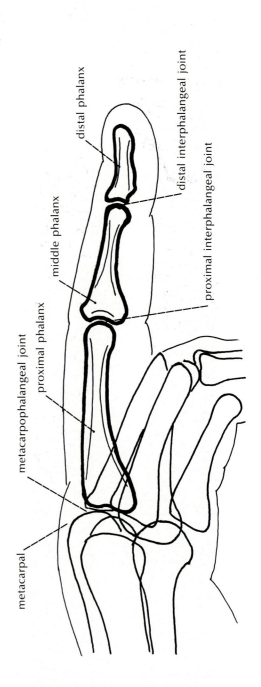

distal phalanx

distal interphalangeal joint

middle phalanx

proximal interphalangeal joint

metacarpophalangeal joint

proximal phalanx

metacarpal

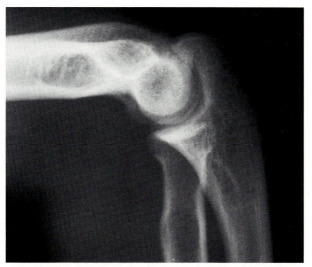

1-11 Elbow, lateral view.

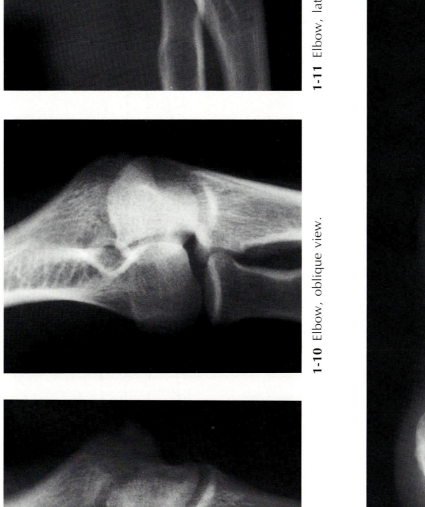

1-10 Elbow, oblique view.

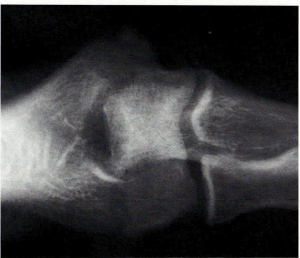

1-9 Elbow, AP view.

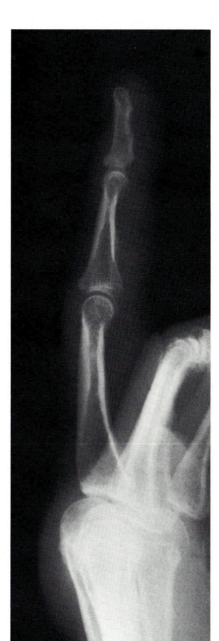

1-12 Finger, lateral view.

1/Forearm

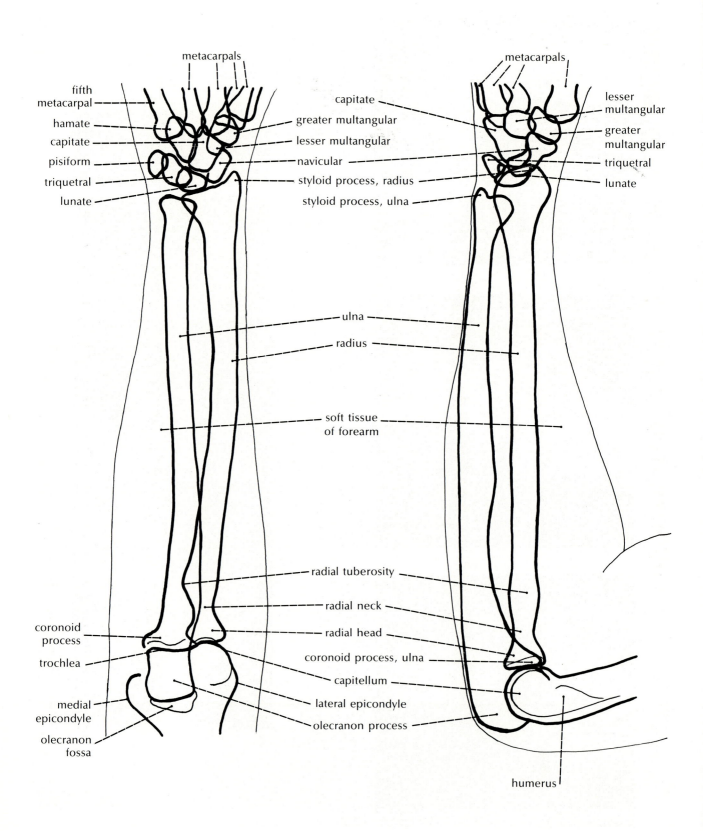

metacarpals

metacarpals

fifth
metacarpal

hamate

capitate

pisiform

triquetral

lunate

capitate

greater multangular

lesser multangular

navicular

styloid process, radius

styloid process, ulna

lesser
multangular

greater
multangular

triquetral

lunate

ulna

radius

soft tissue
of forearm

radial tuberosity

radial neck

radial head

coronoid process, ulna

capitellum

lateral epicondyle

olecranon process

coronoid
process

trochlea

medial
epicondyle

olecranon
fossa

humerus

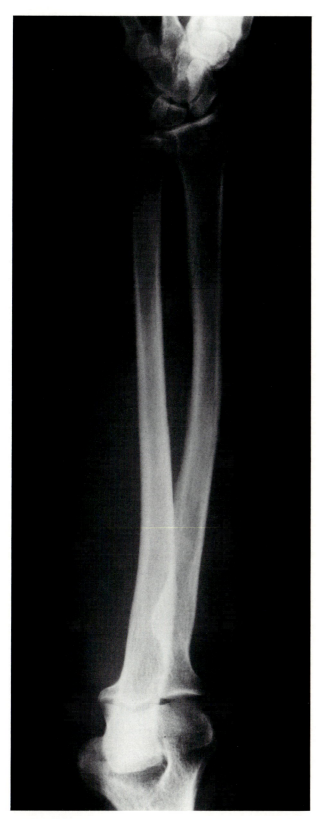

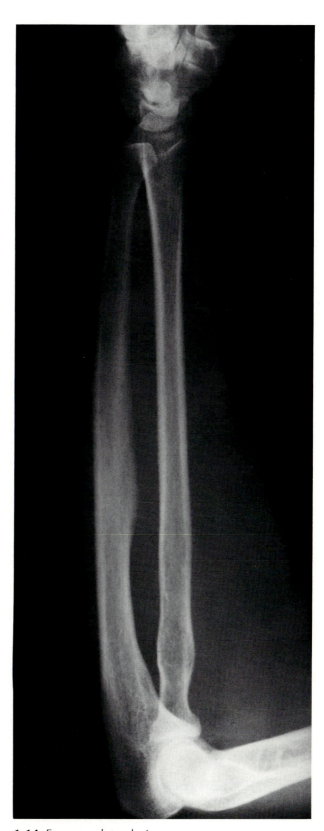

1-13 Forearm, AP view.

1-14 Forearm, lateral view.

1/Forearm and Hand, Arteriography

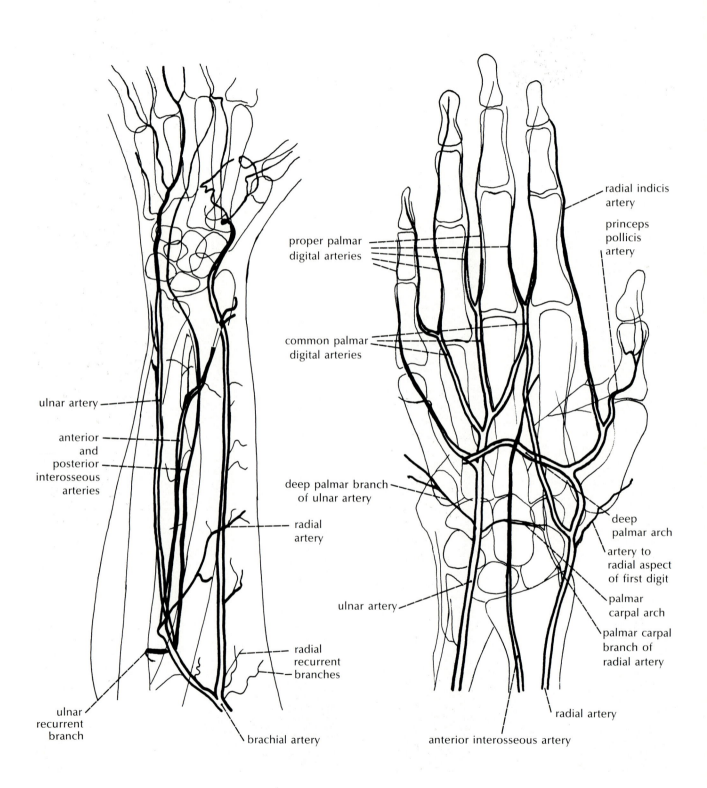

proper palmar
digital arteries

common palmar
digital arteries

ulnar artery

anterior
and
posterior
interosseous
arteries

radial
artery

radial
recurrent
branches

ulnar
recurrent
branch

brachial artery

radial indicis
artery

princeps
pollicis
artery

deep palmar branch
of ulnar artery

ulnar artery

deep
palmar arch

artery to
radial aspect
of first digit

palmar
carpal arch

palmar carpal
branch of
radial artery

radial artery

anterior interosseous artery

1/Forearm and Hand, Arteriography

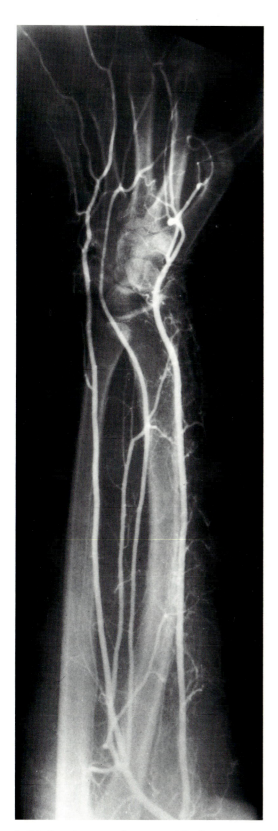

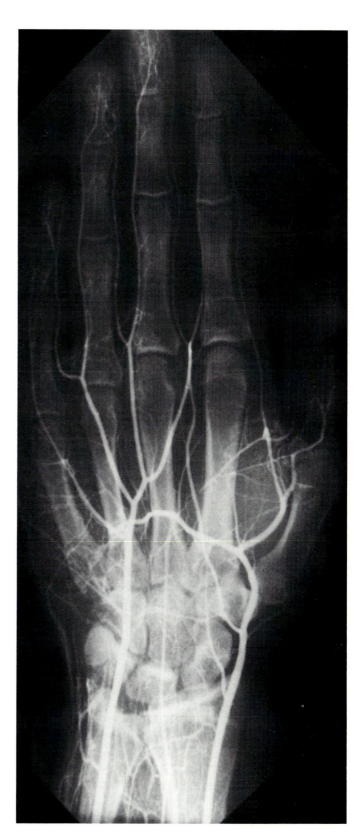

1-15 Arteriogram, forearm, AP view.

1-16 Arteriogram, hand and wrist, AP view.

1/Wrist

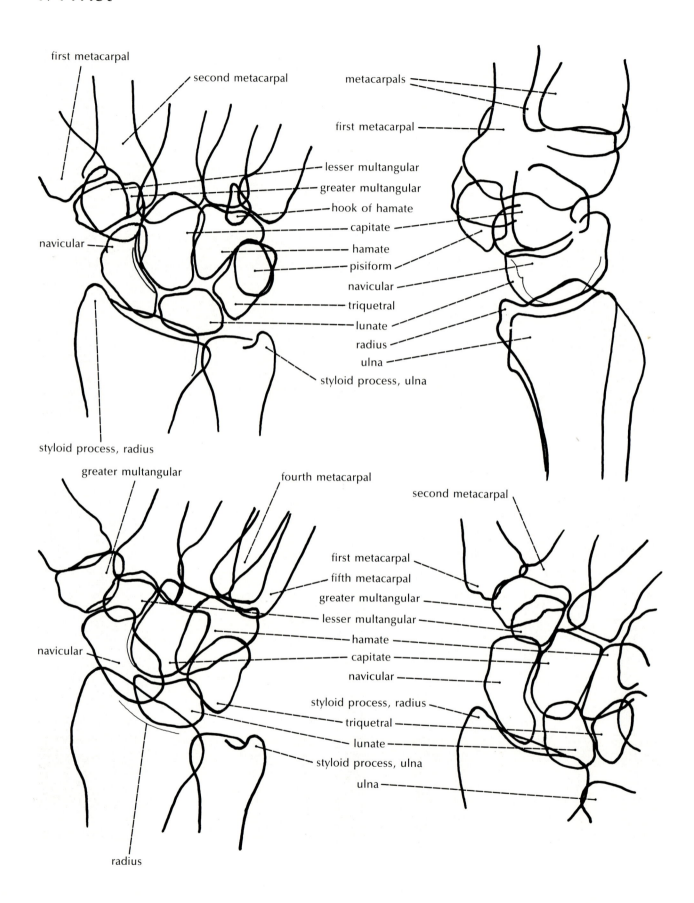

first metacarpal

second metacarpal

metacarpals

first metacarpal

lesser multangular

greater multangular

hook of hamate

capitate

navicular

hamate

pisiform

navicular

triquetral

lunate

radius

ulna

styloid process, ulna

styloid process, radius

greater multangular

fourth metacarpal

second metacarpal

first metacarpal

fifth metacarpal

greater multangular

lesser multangular

hamate

navicular

capitate

navicular

styloid process, radius

triquetral

lunate

styloid process, ulna

ulna

radius

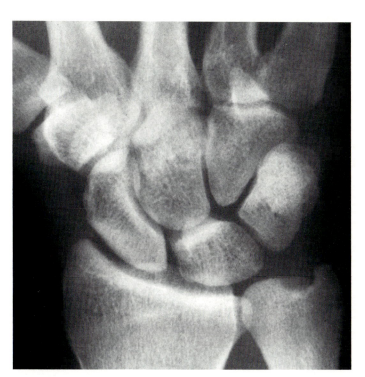

1-17 Wrist, AP view.

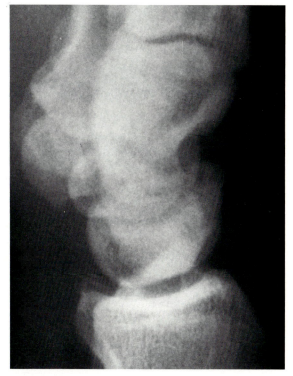

1-18 Wrist, lateral view.

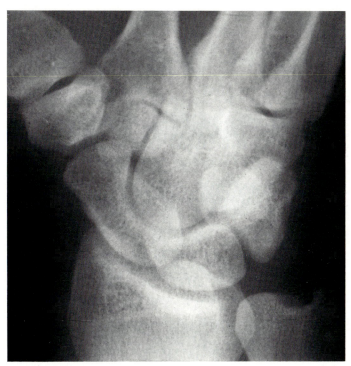

1-19 Wrist, oblique view.

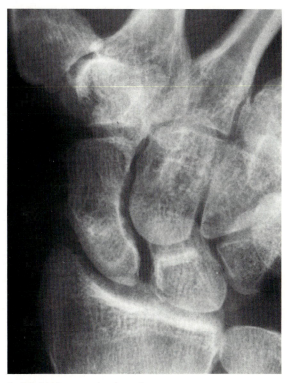

1-20 Wrist, navicular view.

1/Hand and Wrist

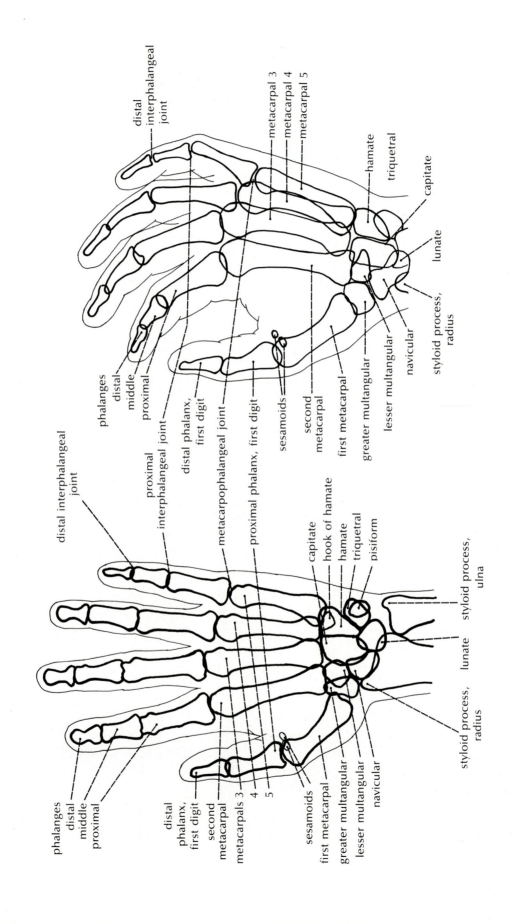

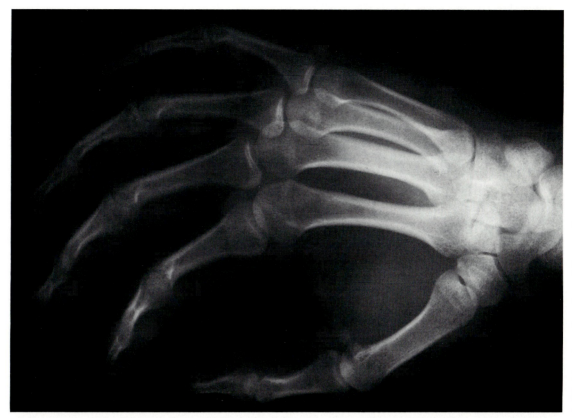

1-22 Hand and wrist, oblique view.

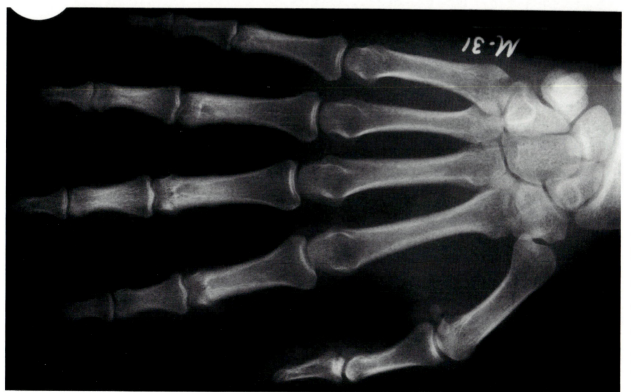

1-21 Hand and wrist, AP view.

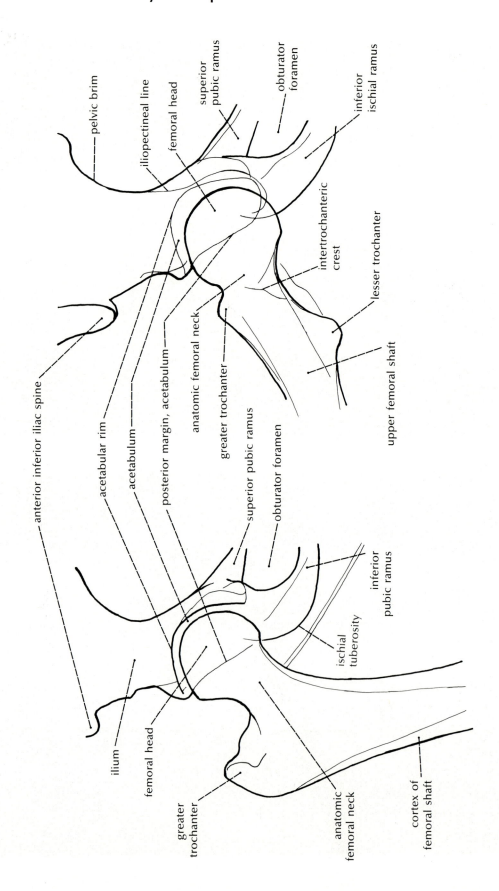

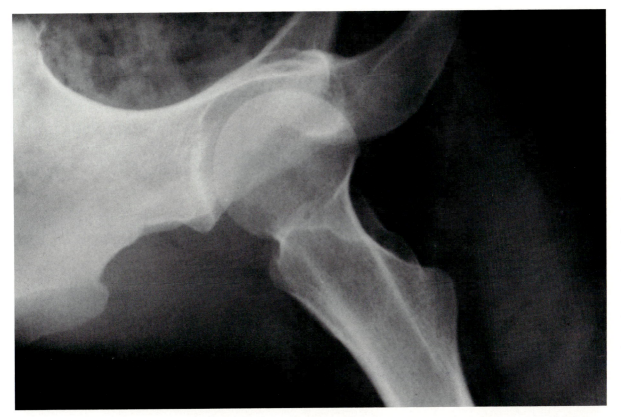

2-2 Hip, lateral view, leg abducted.

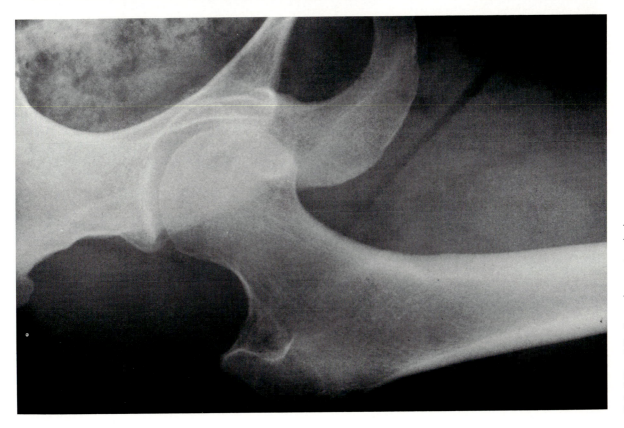

2-1 Hip, AP view, leg extended.

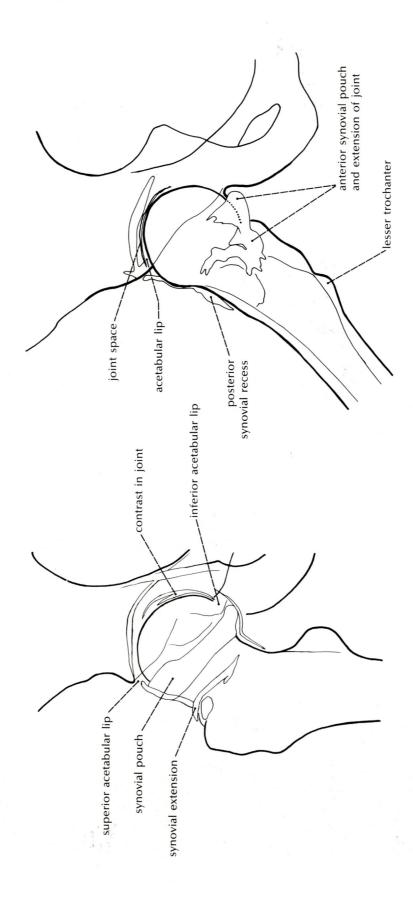

2/Hip, Arthrography

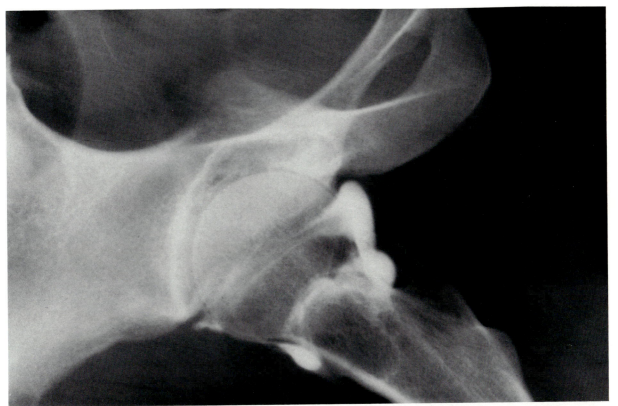

2-4 Hip, double-contrast arthrogram, lateral view.

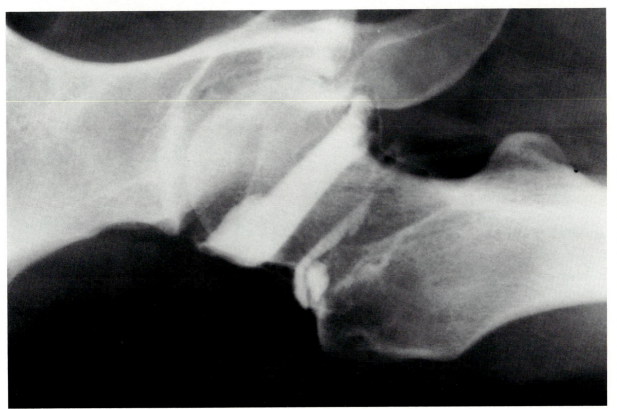

2-3 Hip, double-contrast arthrogram, AP view.

2/Upper Leg

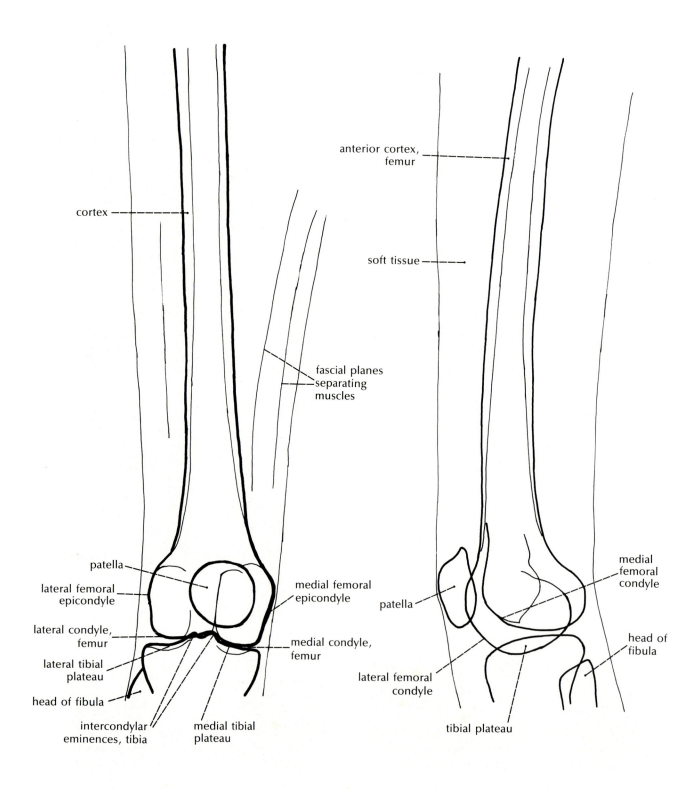

cortex

anterior cortex, femur

soft tissue

fascial planes separating muscles

patella

lateral femoral epicondyle

medial femoral epicondyle

medial femoral condyle

patella

lateral condyle, femur

medial condyle, femur

head of fibula

lateral tibial plateau

lateral femoral condyle

head of fibula

head of fibula

intercondylar eminences, tibia

medial tibial plateau

tibial plateau

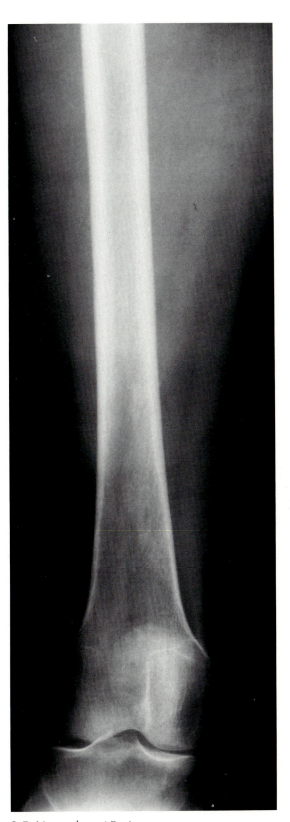

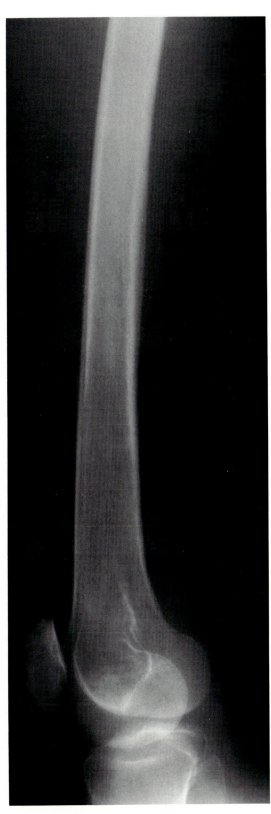

2-5 Upper leg, AP view.

2-6 Upper leg, lateral view.

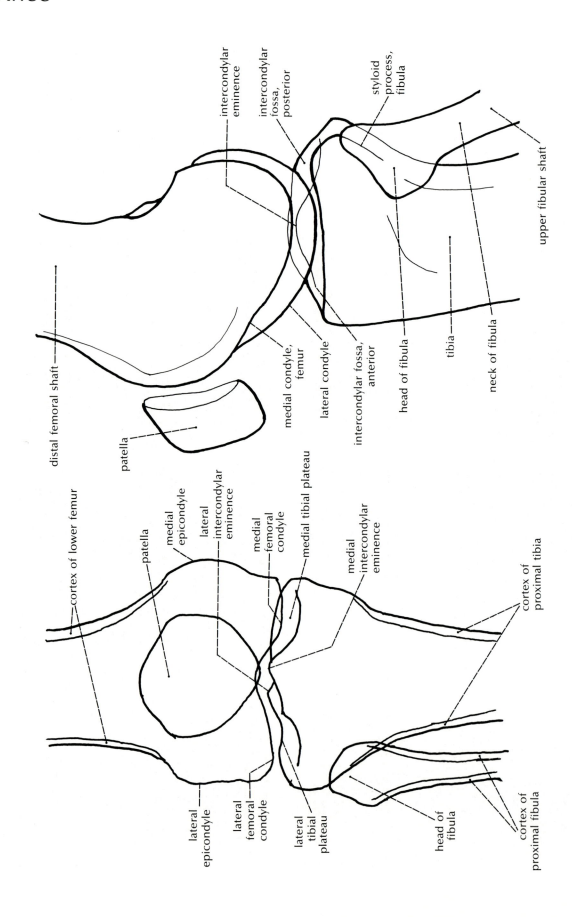

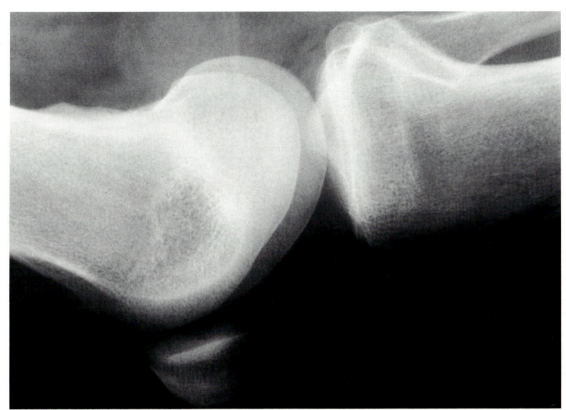

2-8 Knee, lateral view.

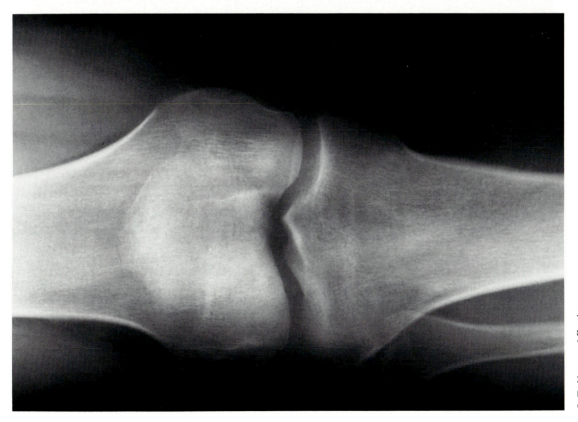

2-7 Knee, AP view.

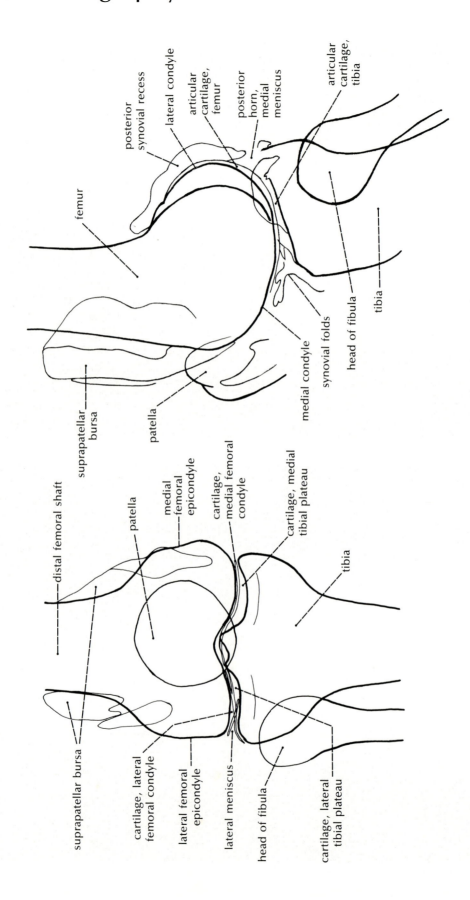

2/Knee, Arthrography

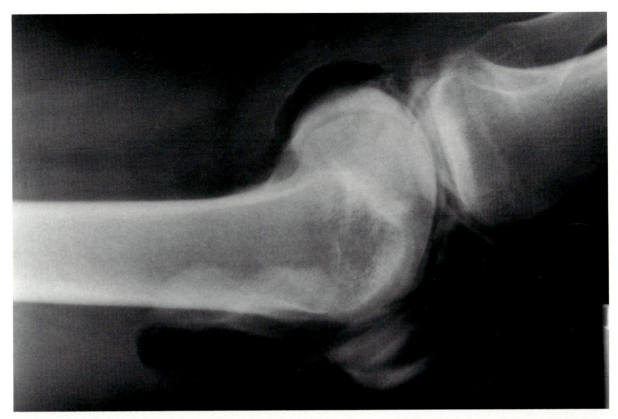

2-10 Knee, double-contrast arthrogram, lateral view.

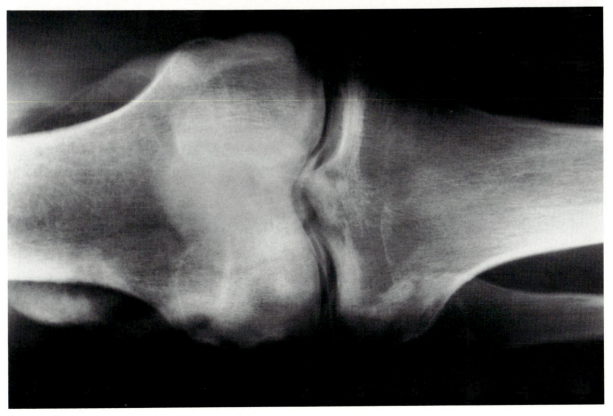

2-9 Knee, double-contrast arthrogram, AP view.

2/Upper Leg, Arteriography

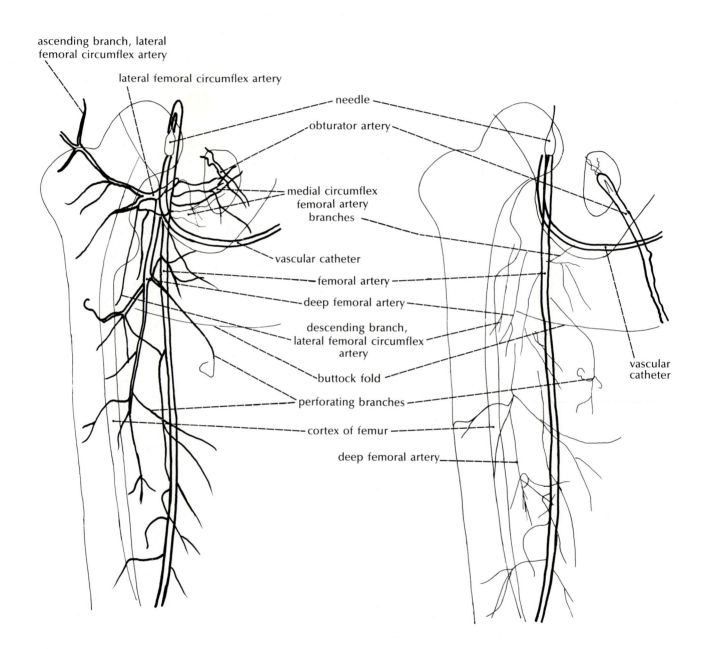

ascending branch, lateral femoral circumflex artery

lateral femoral circumflex artery

needle

obturator artery

medial circumflex femoral artery branches

vascular catheter

femoral artery

deep femoral artery

descending branch, lateral femoral circumflex artery

buttock fold

perforating branches

cortex of femur

deep femoral artery

vascular catheter

2/Upper Leg, Arteriography

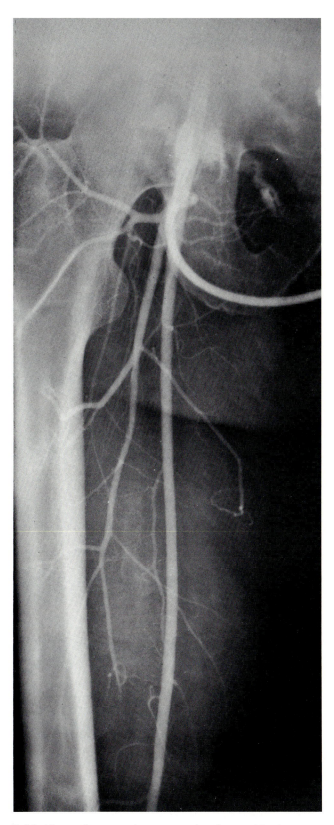

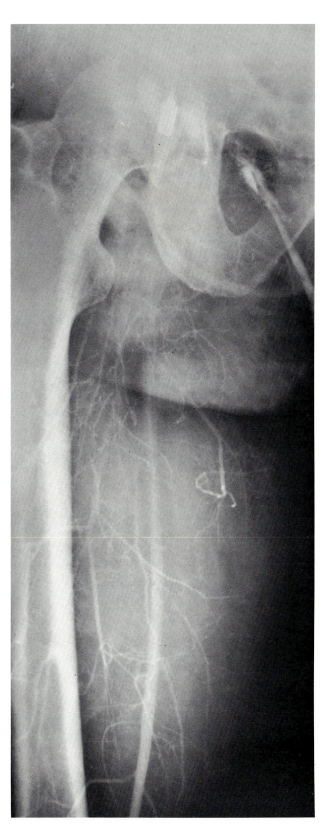

2-11 Upper leg, arteriogram early phase, AP view.

2-12 Upper leg, arteriogram later phase, AP view.

2/Lower Leg

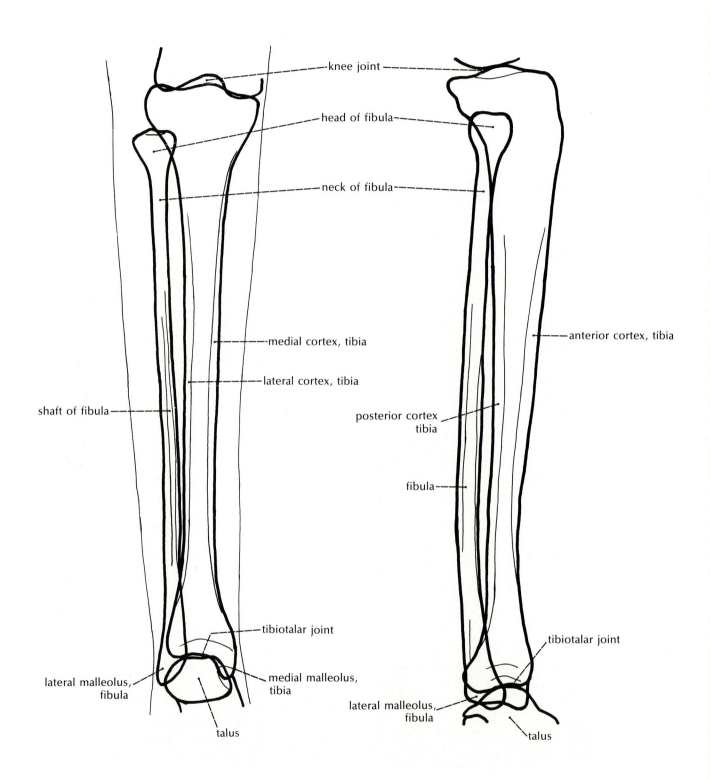

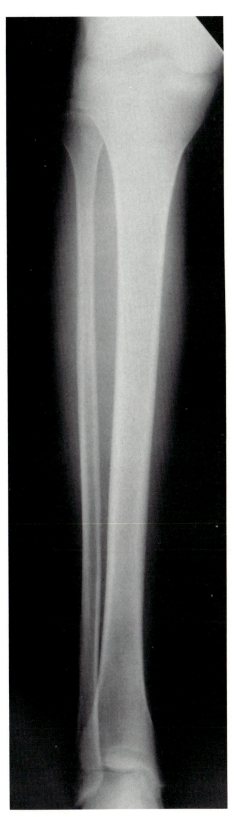

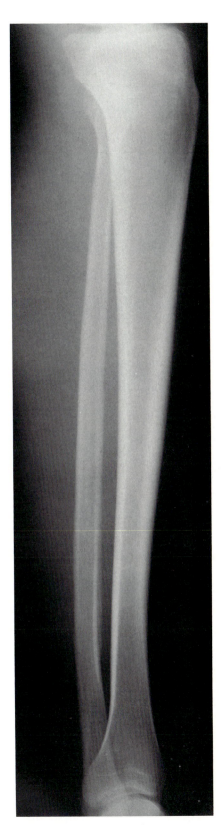

2-13 Lower leg, AP view.

2-14 Lower leg, lateral view.

2/Leg, Venography

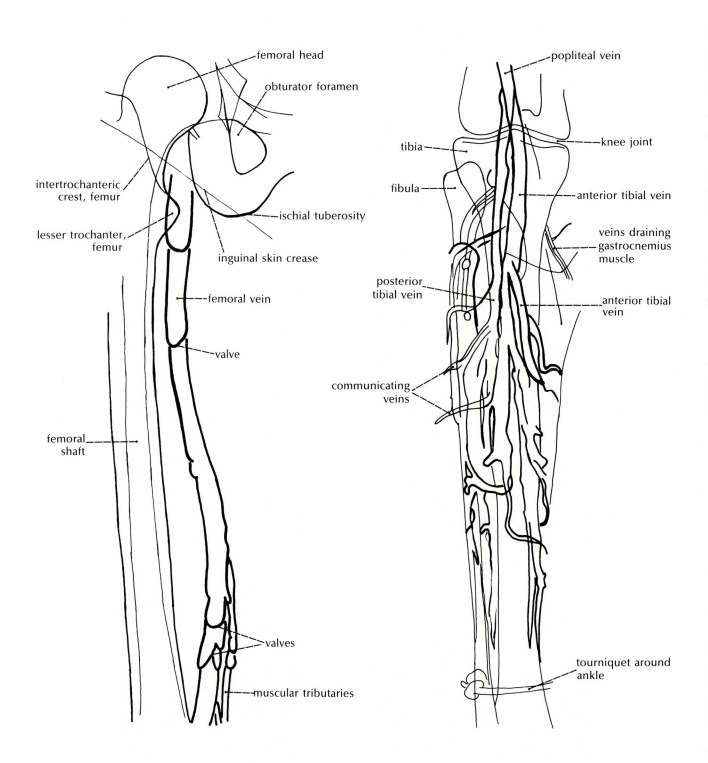

femoral head

obturator foramen

intertrochanteric crest, femur

ischial tuberosity

lesser trochanter, femur

inguinal skin crease

femoral vein

valve

femoral shaft

valves

muscular tributaries

popliteal vein

tibia

knee joint

fibula

anterior tibial vein

veins draining gastrocnemius muscle

posterior tibial vein

anterior tibial vein

communicating veins

tourniquet around ankle

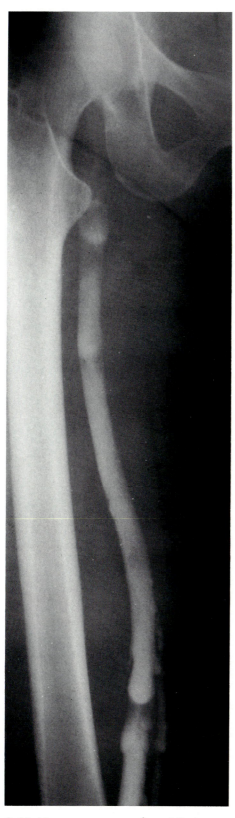

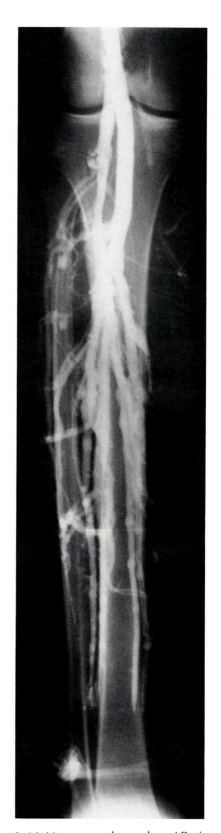

2-15 Venogram, upper leg, AP view.

2-16 Venogram, lower leg, AP view.

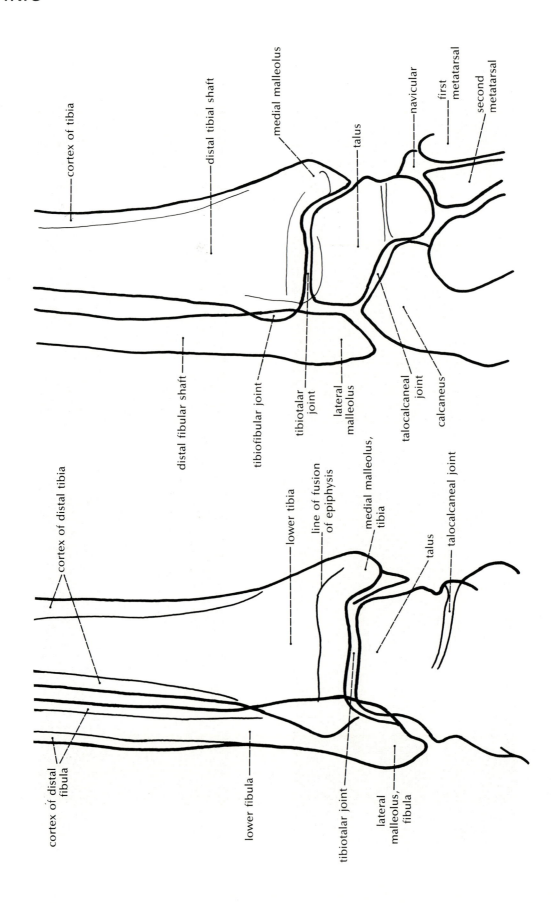

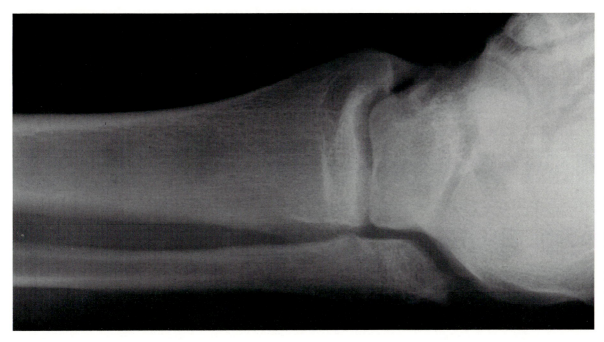

2-18 Ankle, oblique view.

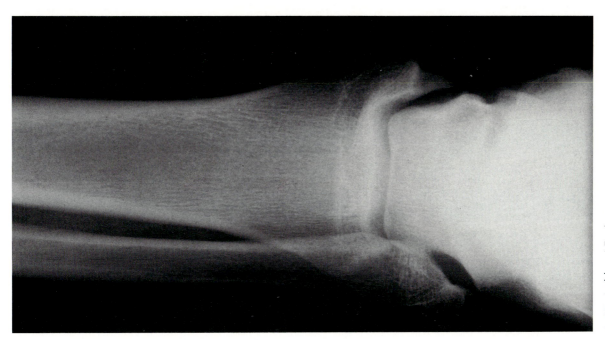

2-17 Ankle, AP view.

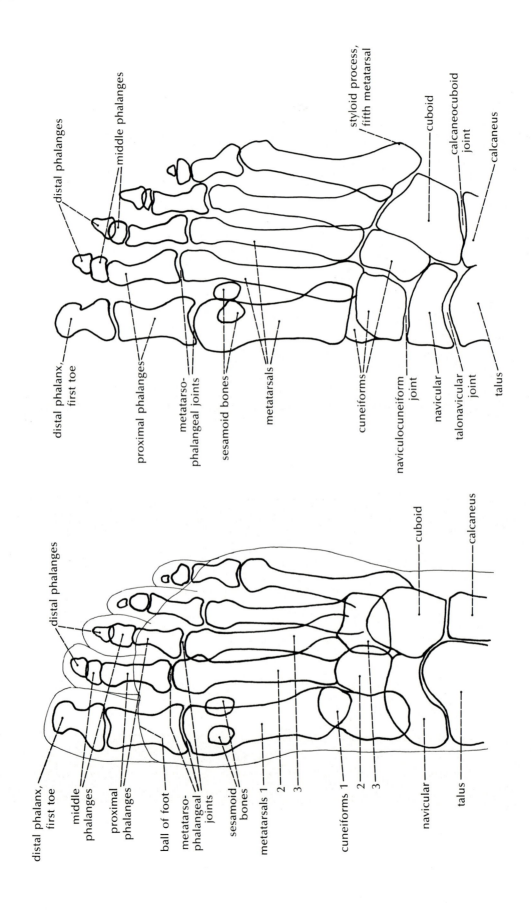

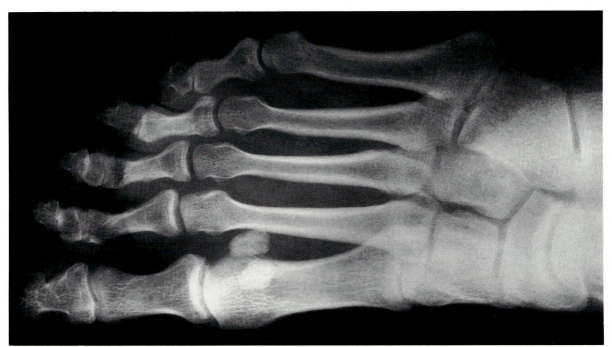

2-20 Foot, oblique view.

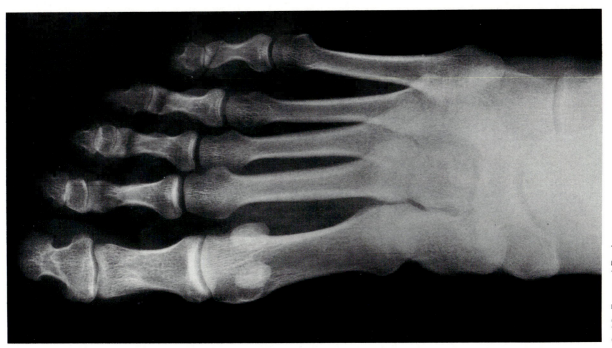

2-19 Foot, AP view.

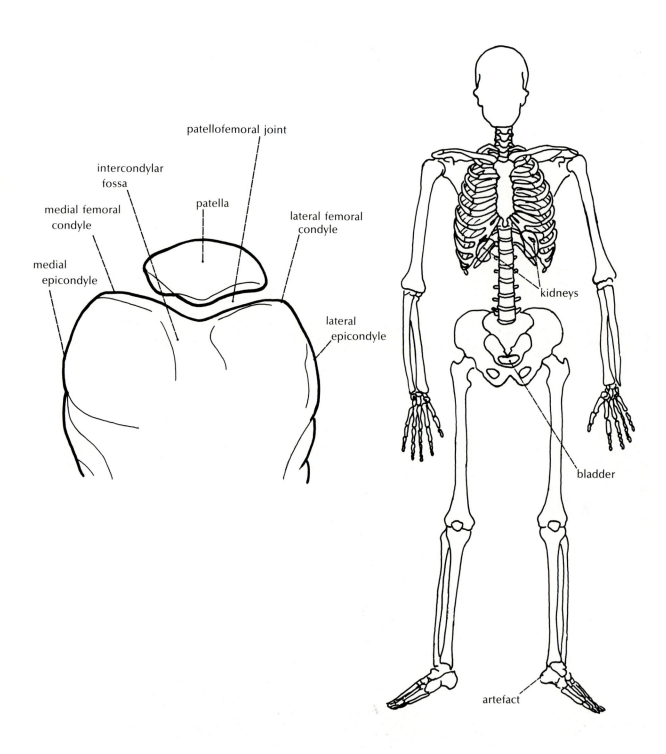

2/Knee (Tangential); Bone Scan

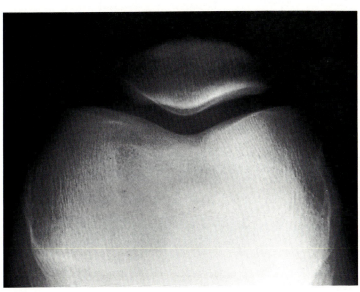

2-21 Knee, tangential, or "sunrise," view.

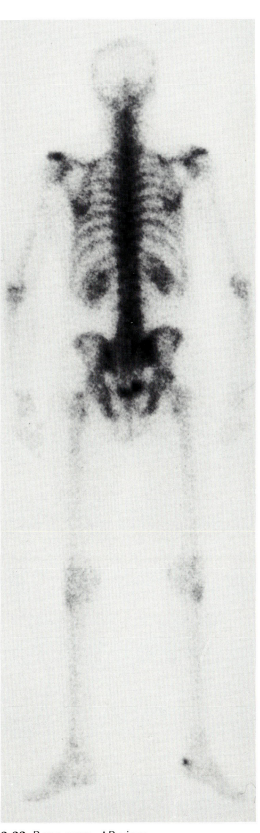

2-22 Bone scan, AP view.

2/Ankle and Foot; Heel

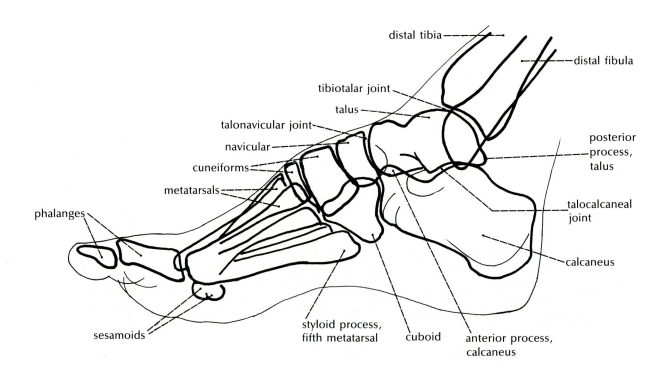

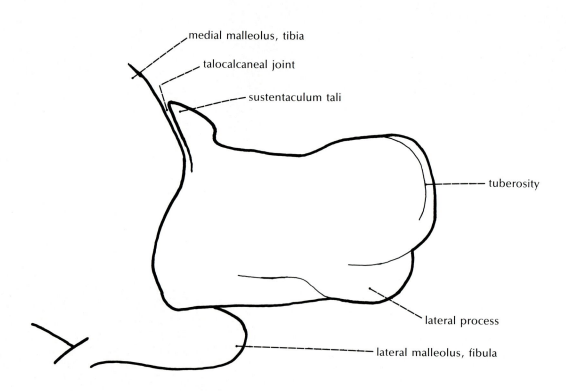

2/Ankle and Foot; Heel

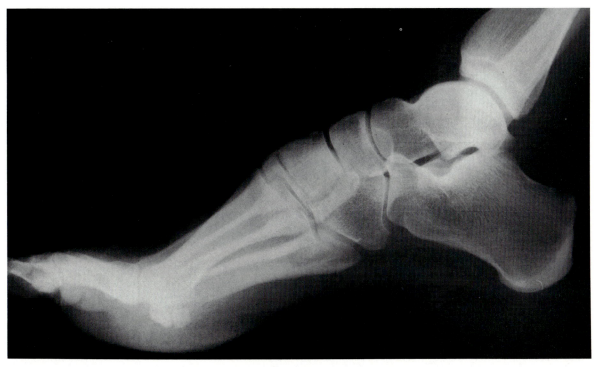

2-23 Ankle and foot, lateral view.

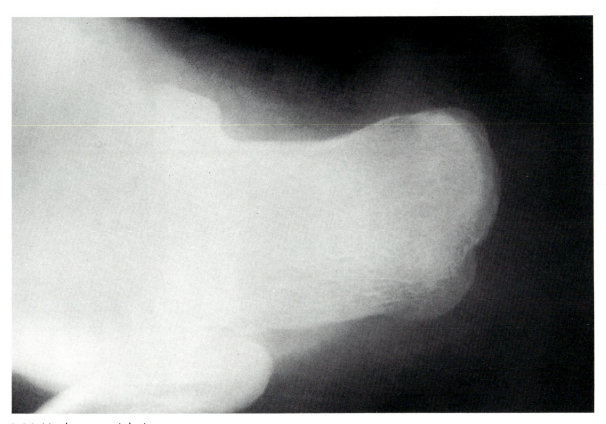

2-24 Heel, tangential view.

3/Skull, Neck, and Spine □ Skull

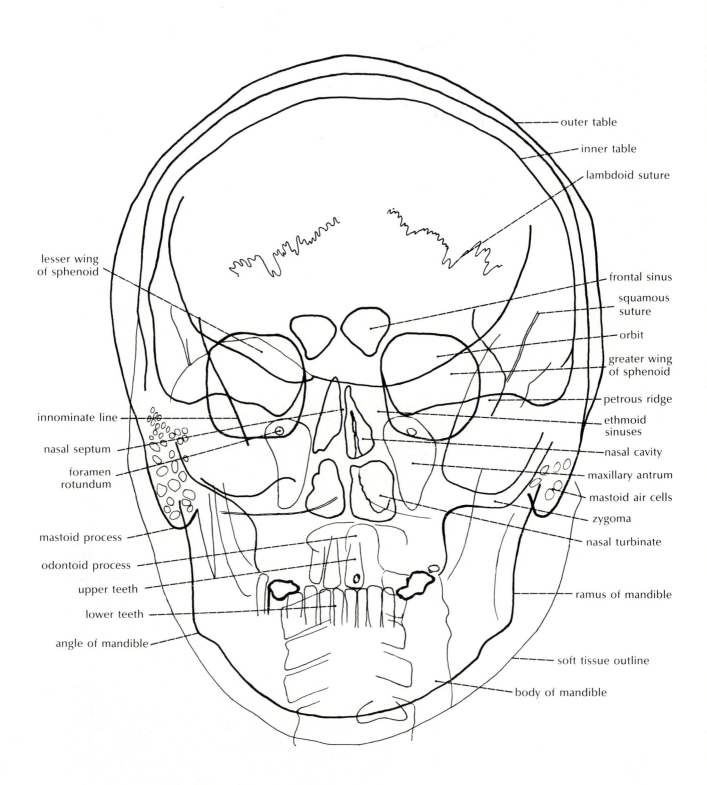

outer table

inner table

lambdoid suture

lesser wing
of sphenoid

frontal sinus

squamous
suture

orbit

greater wing
of sphenoid

petrous ridge

innominate line

ethmoid
sinuses

nasal septum

nasal cavity

foramen
rotundum

maxillary antrum

mastoid air cells

zygoma

mastoid process

nasal turbinate

odontoid process

upper teeth

ramus of mandible

lower teeth

angle of mandible

soft tissue outline

body of mandible

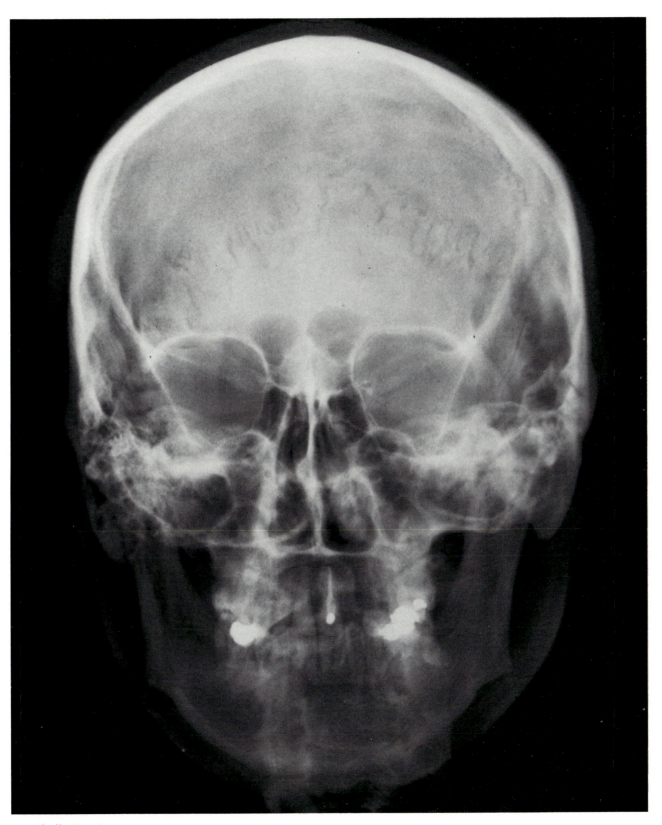

3-1 Skull, PA view.

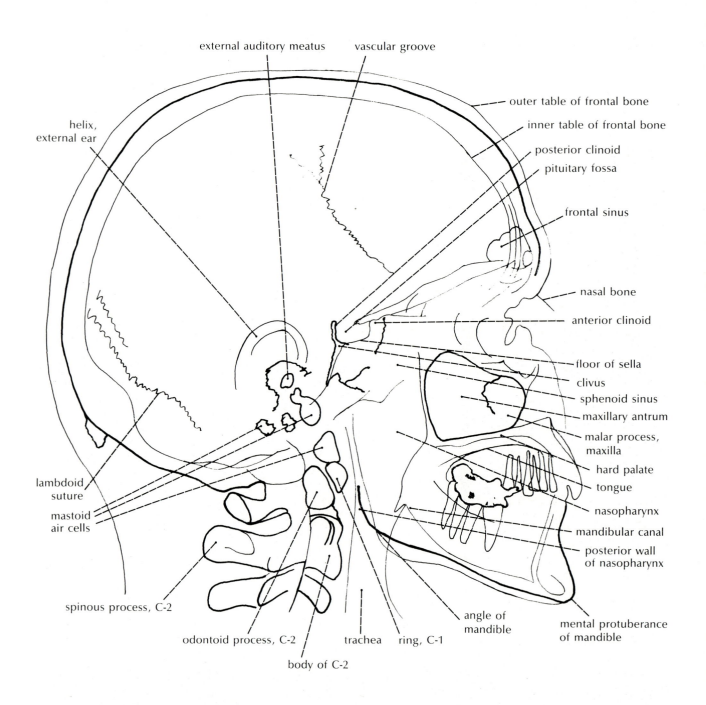

external auditory meatus

vascular groove

helix,
external ear

outer table of frontal bone

inner table of frontal bone

posterior clinoid

pituitary fossa

frontal sinus

nasal bone

anterior clinoid

floor of sella

clivus

sphenoid sinus

maxillary antrum

malar process,
maxilla

hard palate

tongue

nasopharynx

mandibular canal

posterior wall
of nasopharynx

lambdoid
suture

mastoid
air cells

spinous process, C-2

odontoid process, C-2

body of C-2

trachea

ring, C-1

angle of
mandible

mental protuberance
of mandible

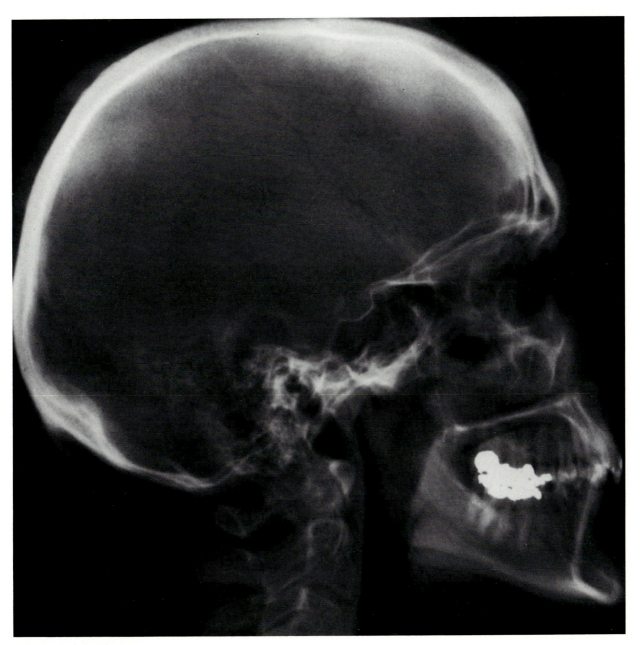

3-2 Skull, lateral view.

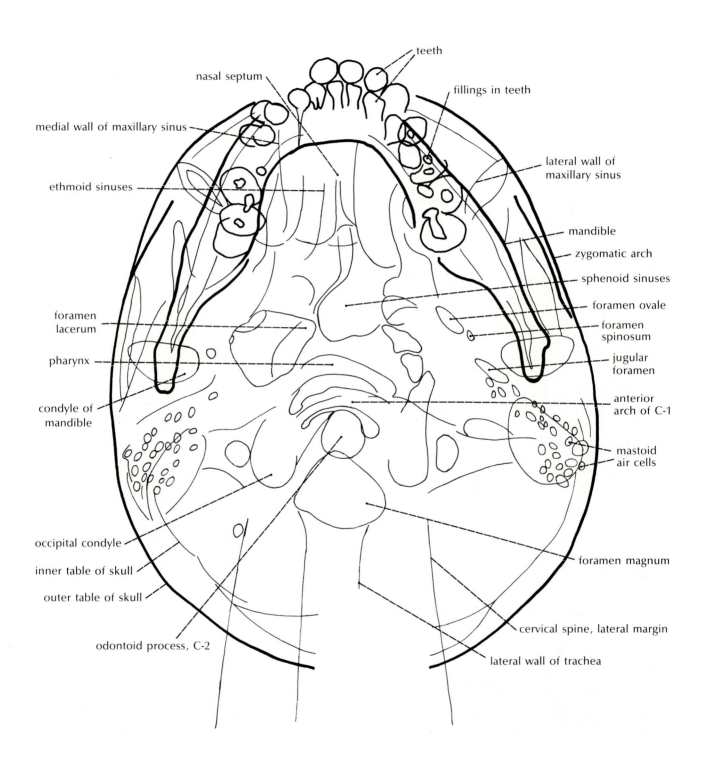

teeth

nasal septum

fillings in teeth

medial wall of maxillary sinus

lateral wall of maxillary sinus

ethmoid sinuses

mandible

zygomatic arch

sphenoid sinuses

foramen ovale

foramen lacerum

foramen spinosum

pharynx

jugular foramen

anterior arch of C-1

condyle of mandible

mastoid air cells

occipital condyle

inner table of skull

outer table of skull

foramen magnum

odontoid process, C-2

cervical spine, lateral margin

lateral wall of trachea

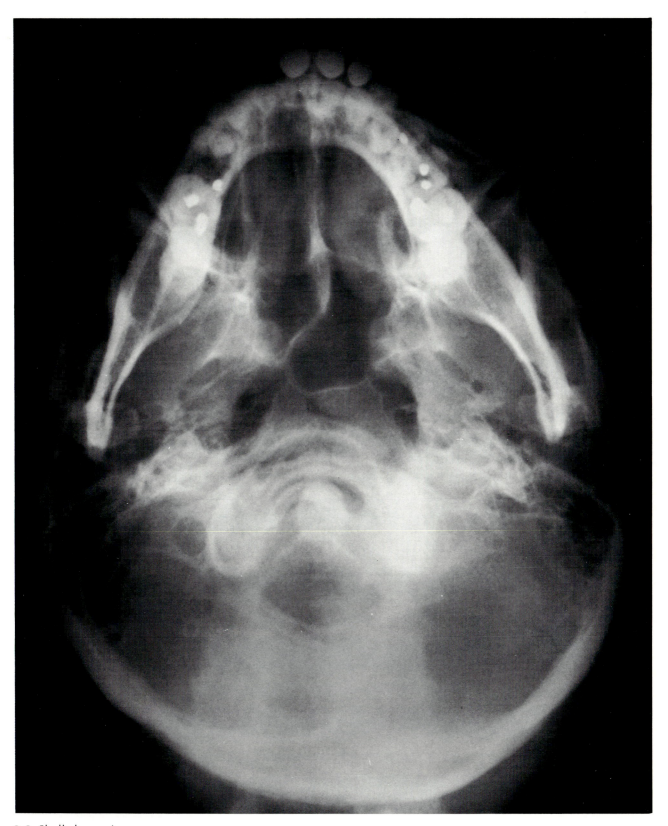

3-3 Skull, base view.

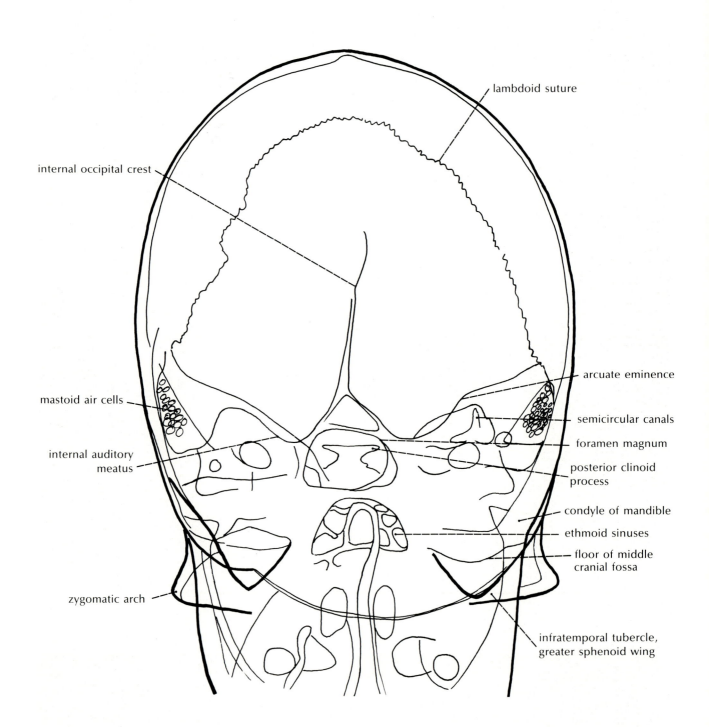

lambdoid suture

internal occipital crest

arcuate eminence

mastoid air cells

semicircular canals

foramen magnum

internal auditory
meatus

posterior clinoid
process

condyle of mandible

ethmoid sinuses

floor of middle
cranial fossa

zygomatic arch

infratemporal tubercle,
greater sphenoid wing

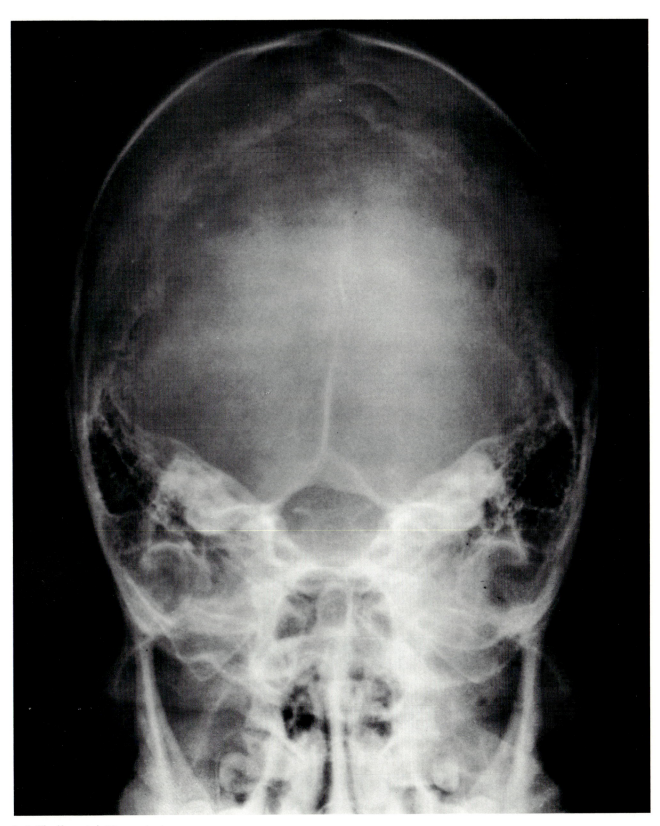

3-4 Skull, axial (Towne) view.

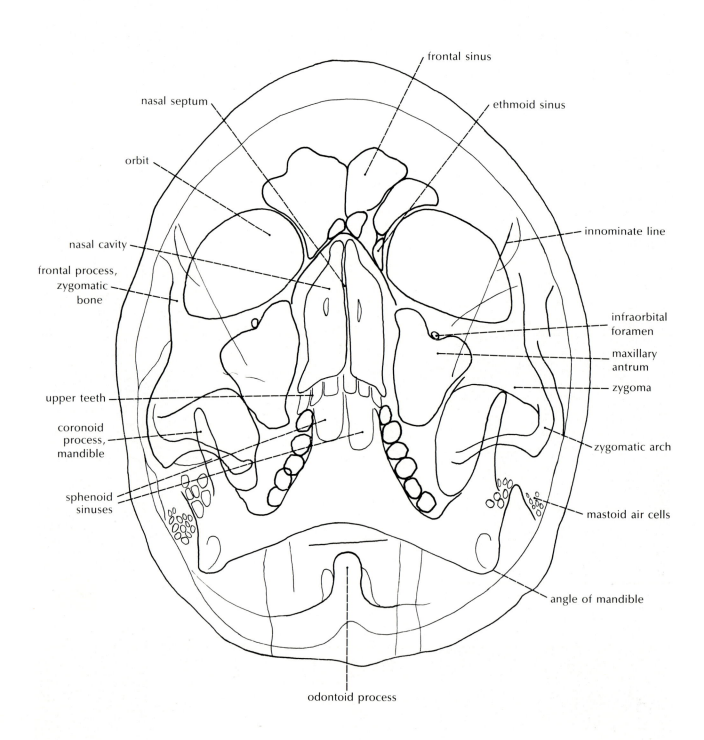

frontal sinus

nasal septum

ethmoid sinus

orbit

innominate line

nasal cavity

frontal process,
zygomatic
bone

infraorbital
foramen

maxillary
antrum

zygoma

upper teeth

coronoid
process,
mandible

zygomatic arch

sphenoid
sinuses

mastoid air cells

angle of mandible

odontoid process

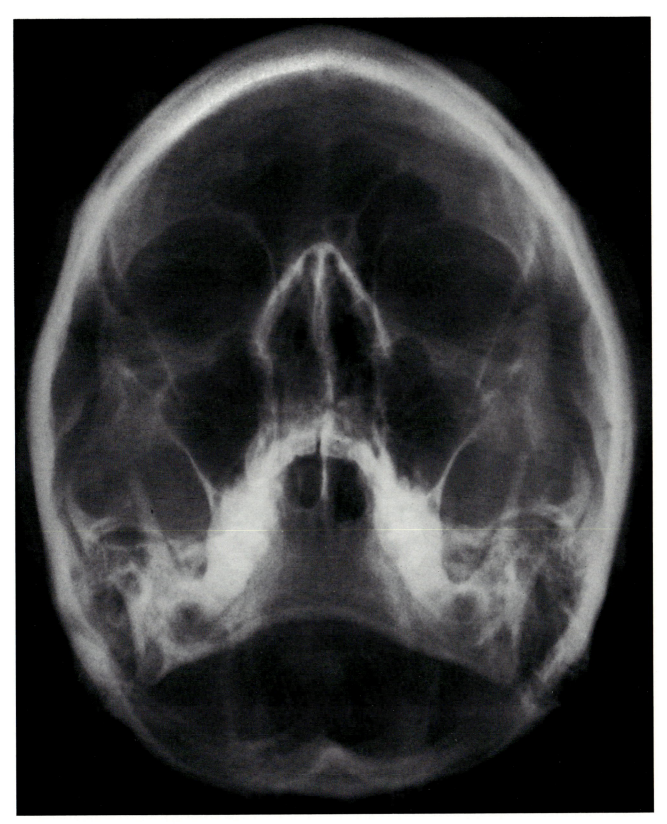

3-5 Skull, frontal view angled (Waters view).

3/Mastoid

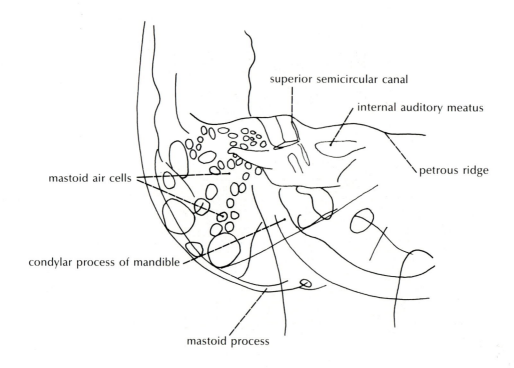

superior semicircular canal

internal auditory meatus

petrous ridge

mastoid air cells

condylar process of mandible

mastoid process

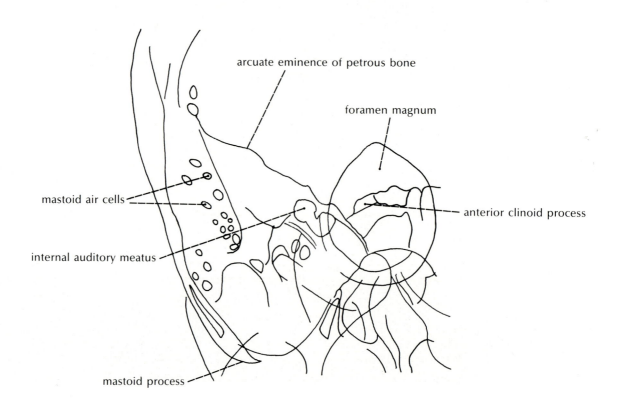

arcuate eminence of petrous bone

foramen magnum

mastoid air cells

anterior clinoid process

internal auditory meatus

mastoid process

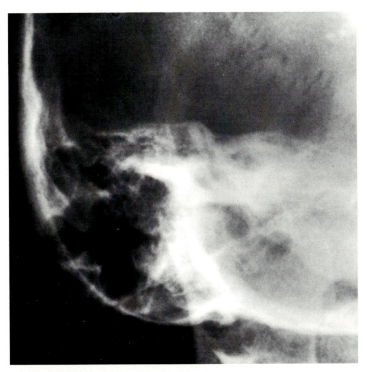

3-6 Mastoid, Stenvers view.

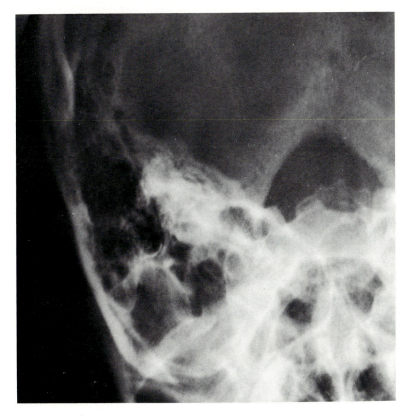

3-7 Mastoid, Towne view.

3/Temporomandibular Joint

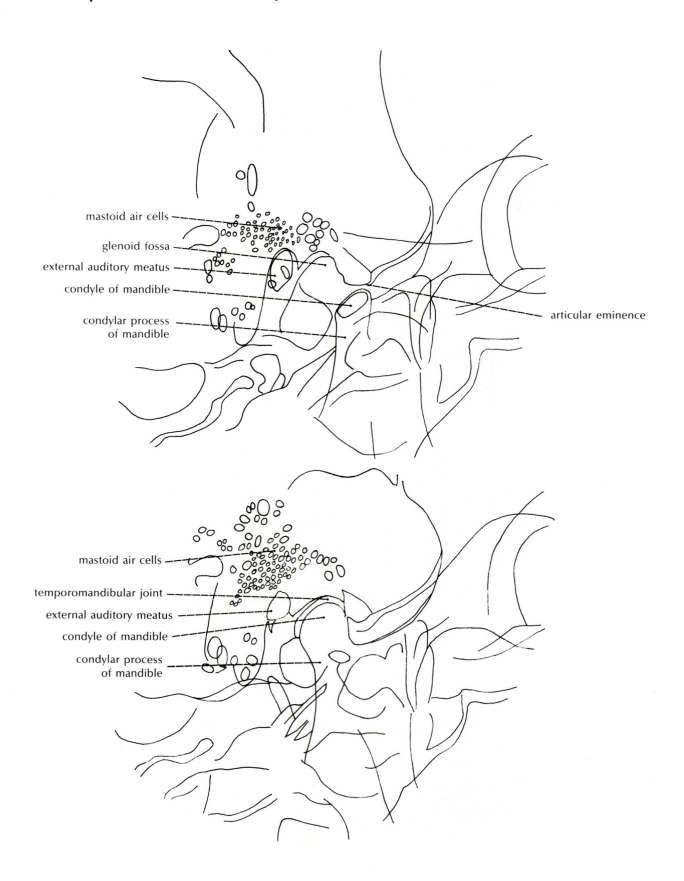

mastoid air cells

glenoid fossa

external auditory meatus

condyle of mandible

condylar process
of mandible

articular eminence

mastoid air cells

temporomandibular joint

external auditory meatus

condyle of mandible

condylar process
of mandible

3/Temporomandibular Joint

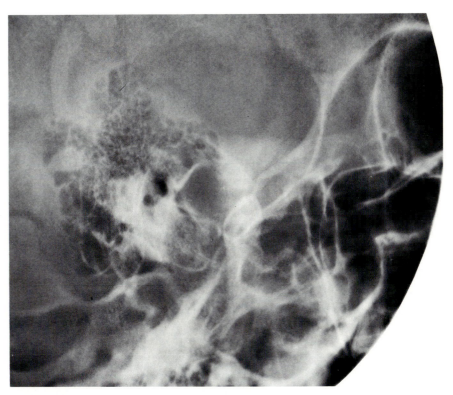

3-8 Temporomandibular joint and mastoid, lateral view (Law), open mouth position.

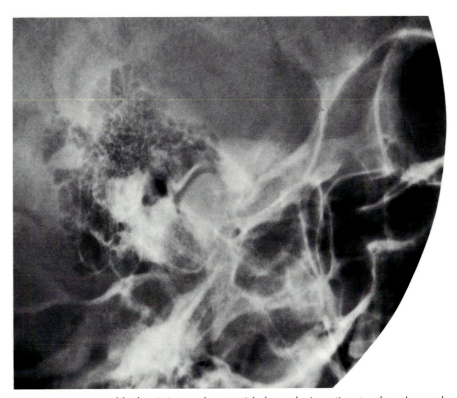

3-9 Temporomandibular joint and mastoid, lateral view (Law), closed mouth position.

3/Sella Turcica; Optic Foramen

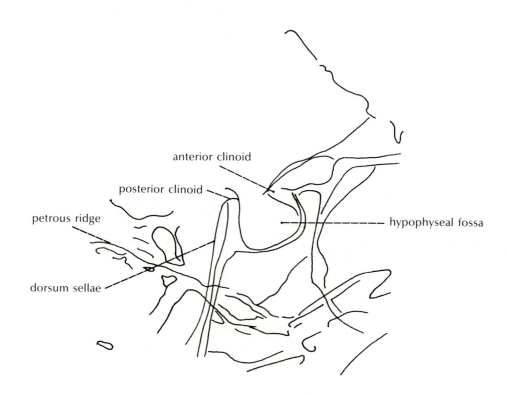

anterior clinoid

posterior clinoid

petrous ridge

hypophyseal fossa

dorsum sellae

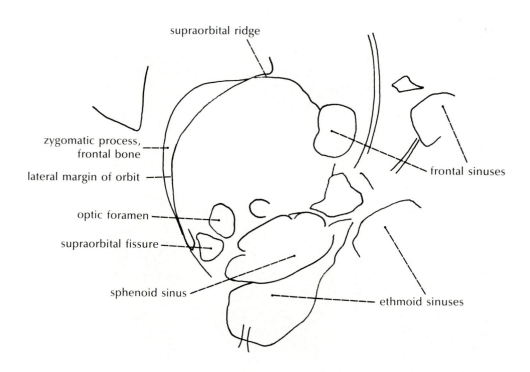

supraorbital ridge

zygomatic process, frontal bone

lateral margin of orbit

optic foramen

supraorbital fissure

sphenoid sinus

frontal sinuses

ethmoid sinuses

3/Sella Turcica; Optic Foramen

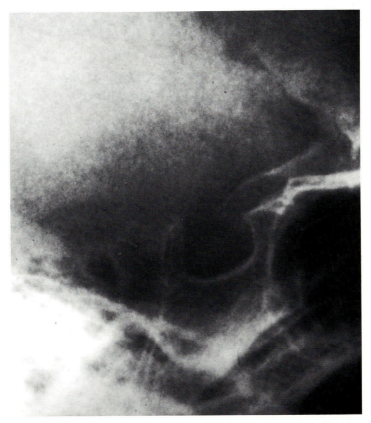

3-10 Sella turcica, lateral view.

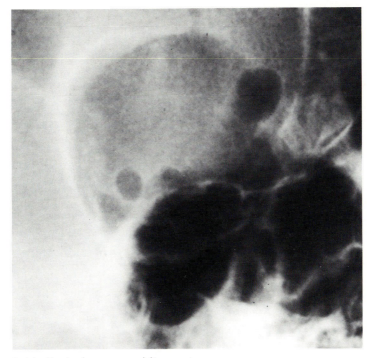

3-11 Optic foramen, oblique view.

3/Cervical Spine

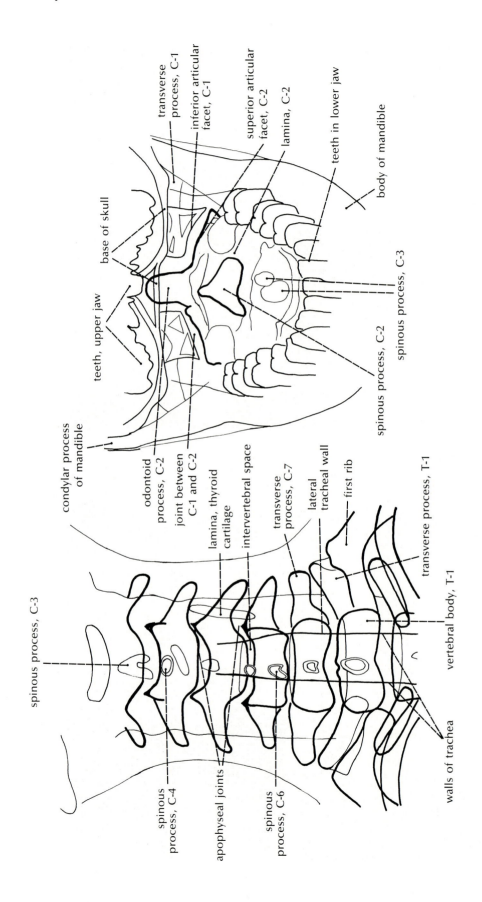

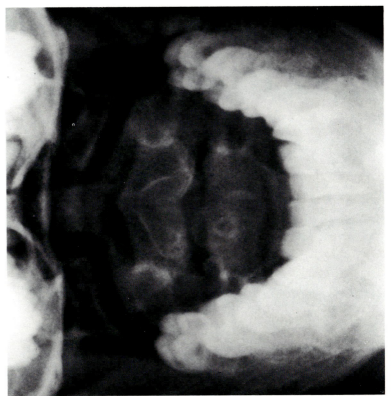

3-13 Cervical spine, open mouth view.

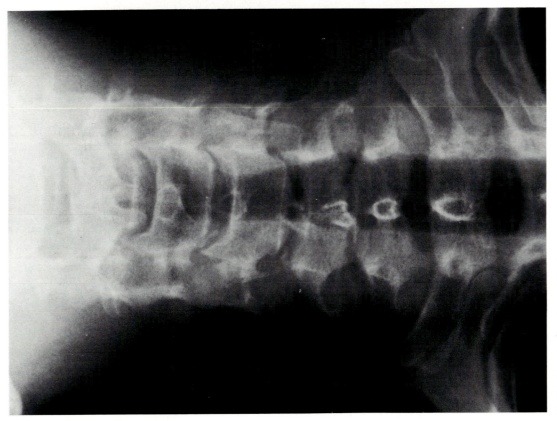

3-12 Cervical spine, AP view.

3/Cervical Spine

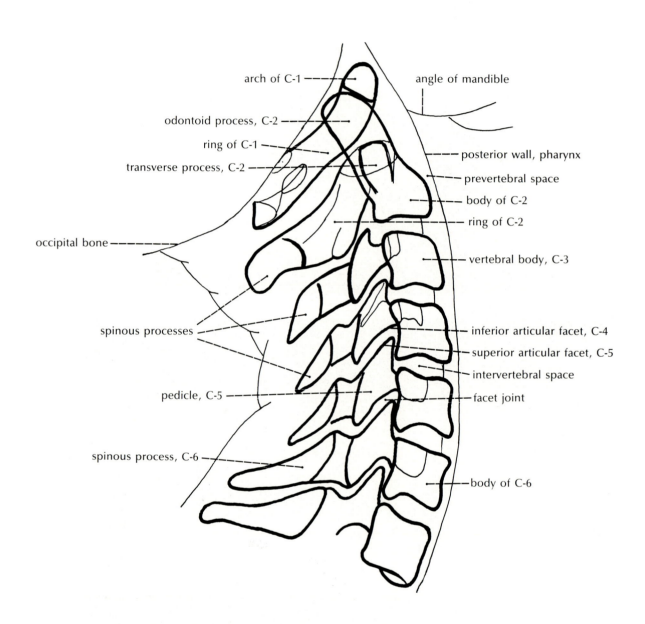

arch of C-1

odontoid process, C-2

ring of C-1

transverse process, C-2

occipital bone

spinous processes

pedicle, C-5

spinous process, C-6

angle of mandible

posterior wall, pharynx

prevertebral space

body of C-2

ring of C-2

vertebral body, C-3

inferior articular facet, C-4

superior articular facet, C-5

intervertebral space

facet joint

body of C-6

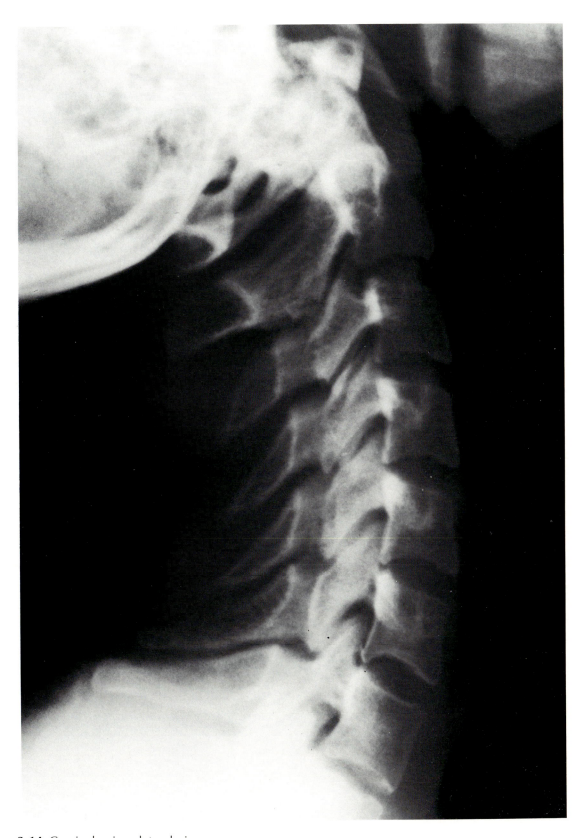

3-14 Cervical spine, lateral view.

3/Cervical Spine

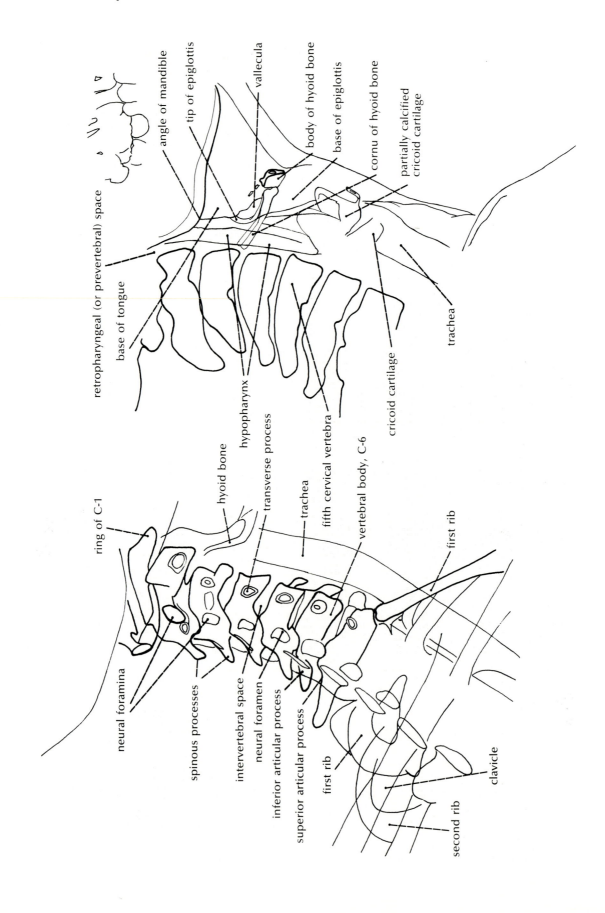

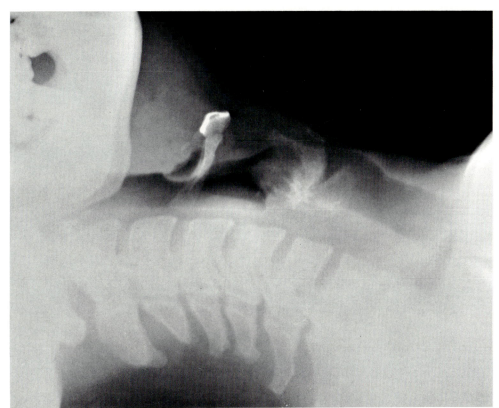

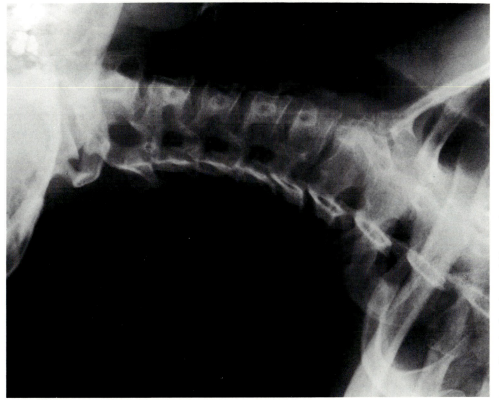

3-16 Soft tissue of the neck, lateral view.

3-15 Cervical spine, oblique view.

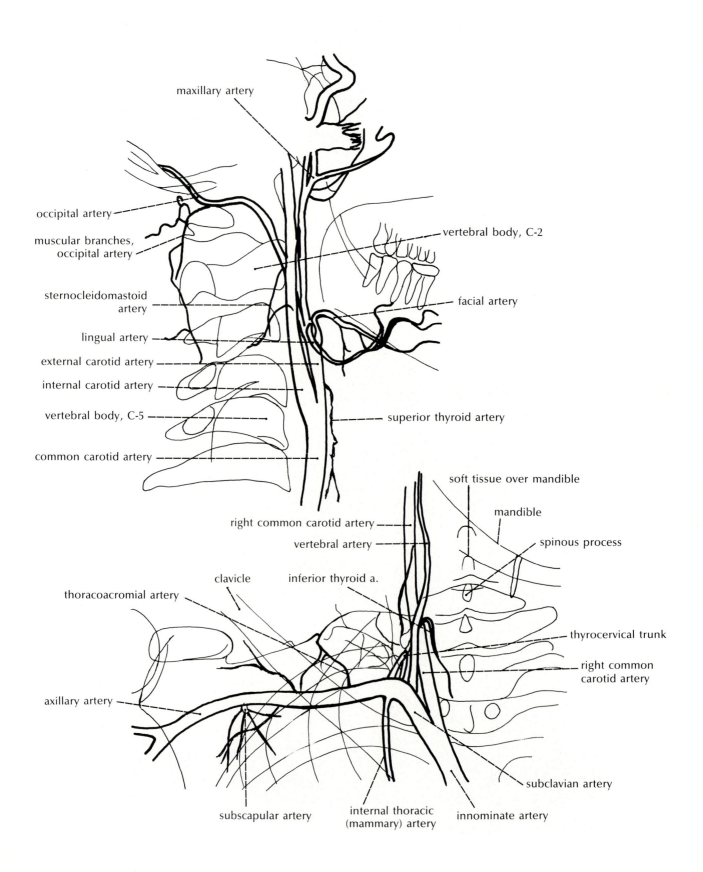

maxillary artery

occipital artery

muscular branches, occipital artery

vertebral body, C-2

sternocleidomastoid artery

facial artery

lingual artery

external carotid artery

internal carotid artery

vertebral body, C-5

superior thyroid artery

common carotid artery

soft tissue over mandible

right common carotid artery

mandible

vertebral artery

spinous process

clavicle

inferior thyroid a.

thoracoacromial artery

thyrocervical trunk

right common carotid artery

axillary artery

subclavian artery

subscapular artery

internal thoracic (mammary) artery

innominate artery

3/Neck, Arteriography

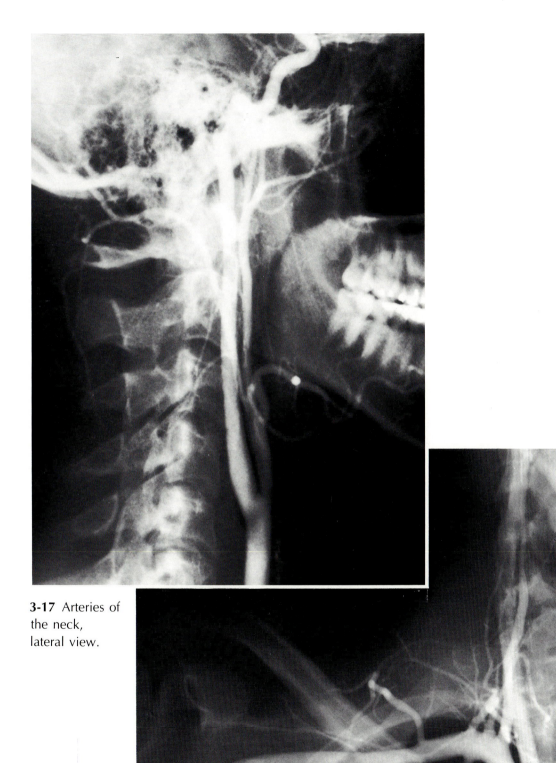

3-17 Arteries of the neck, lateral view.

3-18 Arteries of the neck, AP view.

3/Cervical Esophagus

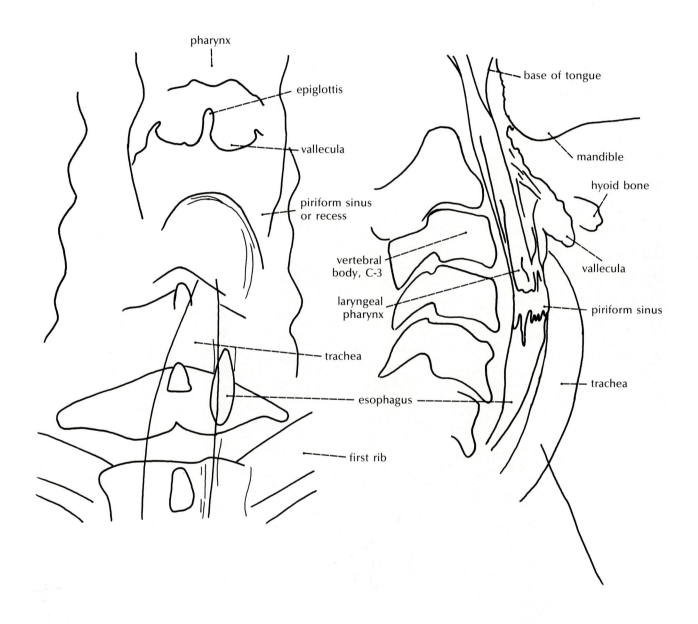

pharynx

epiglottis

vallecula

piriform sinus
or recess

vertebral
body, C-3

laryngeal
pharynx

trachea

esophagus

first rib

base of tongue

mandible

hyoid bone

vallecula

piriform sinus

trachea

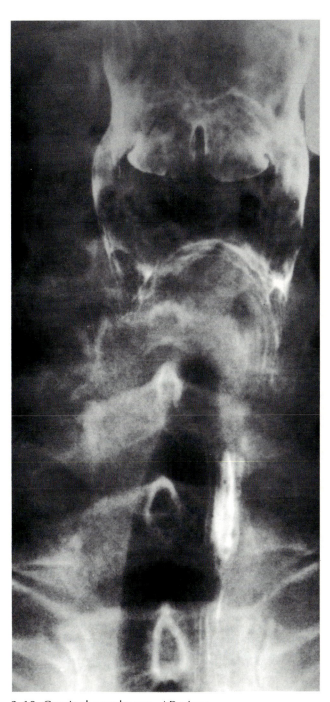

3-19 Cervical esophagus, AP view.

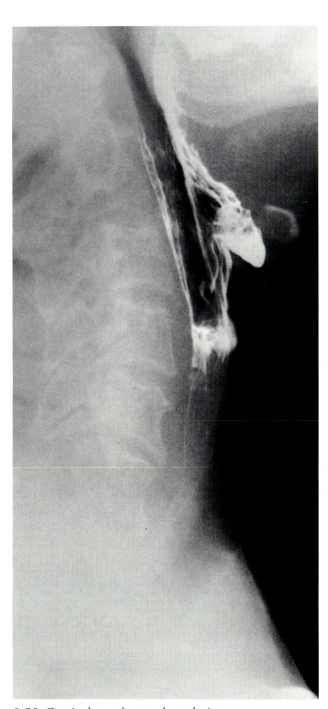

3-20 Cervical esophagus, lateral view.

3/Base of Skull and Upper Neck, Computed Tomography

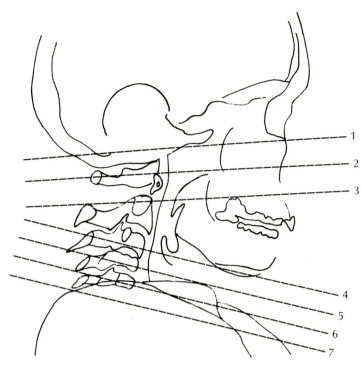

3-21 Cross sections of the base of the skull and neck recorded by the CT scanner. Sections 1–3 are of the base of the skull and the upper neck. Sections 4–7 are of the mid and lower neck at a different angle from a different subject.

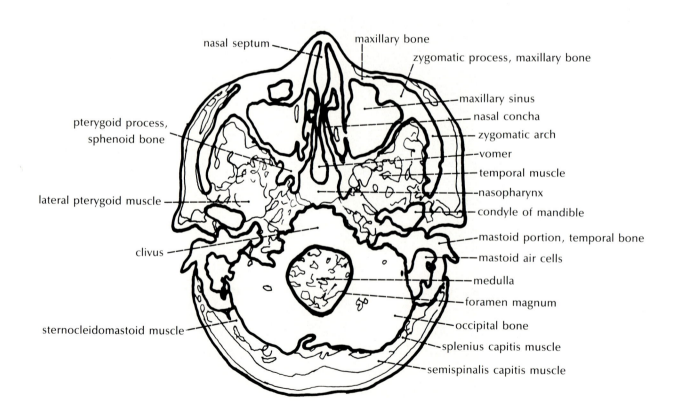

nasal septum — maxillary bone

zygomatic process, maxillary bone

maxillary sinus

pterygoid process, sphenoid bone

nasal concha

zygomatic arch

vomer

temporal muscle

lateral pterygoid muscle

nasopharynx

condyle of mandible

mastoid portion, temporal bone

clivus

mastoid air cells

medulla

foramen magnum

sternocleidomastoid muscle

occipital bone

splenius capitis muscle

semispinalis capitis muscle

3/Base of Skull and Upper Neck, Computed Tomography

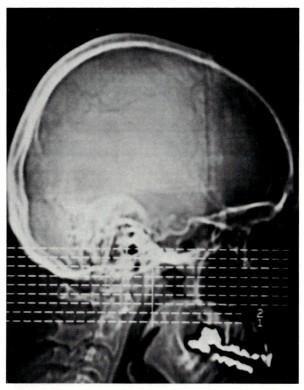

3-21 Cross sections.

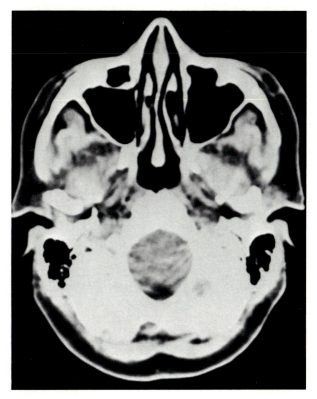

3-22 Section 1.

3/Base of Skull and Upper Neck, Computed Tomography

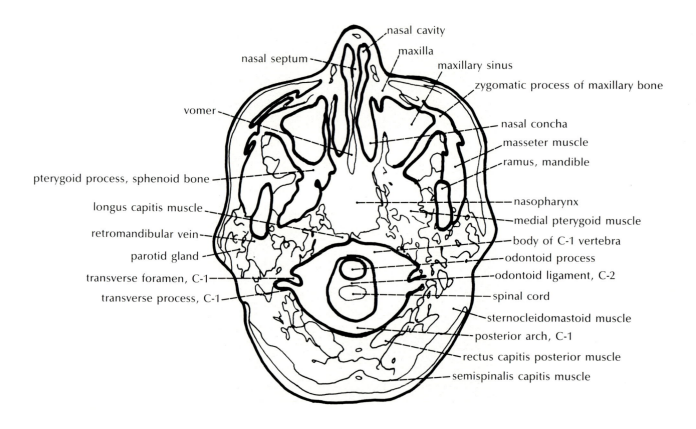

- nasal cavity
- nasal septum
- maxilla
- maxillary sinus
- zygomatic process of maxillary bone
- vomer
- nasal concha
- masseter muscle
- ramus, mandible
- pterygoid process, sphenoid bone
- longus capitis muscle
- nasopharynx
- medial pterygoid muscle
- retromandibular vein
- body of C-1 vertebra
- parotid gland
- odontoid process
- transverse foramen, C-1
- odontoid ligament, C-2
- transverse process, C-1
- spinal cord
- sternocleidomastoid muscle
- posterior arch, C-1
- rectus capitis posterior muscle
- semispinalis capitis muscle

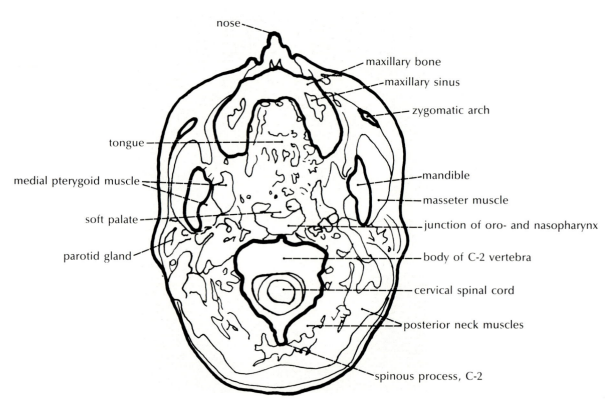

- nose
- maxillary bone
- maxillary sinus
- zygomatic arch
- tongue
- medial pterygoid muscle
- mandible
- masseter muscle
- soft palate
- junction of oro- and nasopharynx
- parotid gland
- body of C-2 vertebra
- cervical spinal cord
- posterior neck muscles
- spinous process, C-2

3/Base of Skull and Upper Neck, Computed Tomography

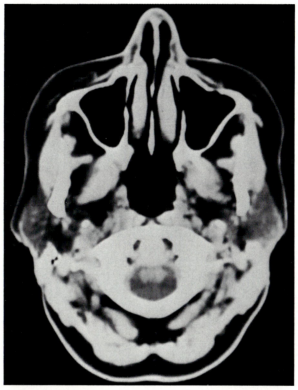

3-23 Section 2.

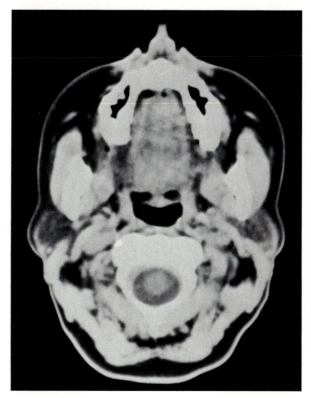

3-24 Section 3.

3/Neck, Computed Tomography

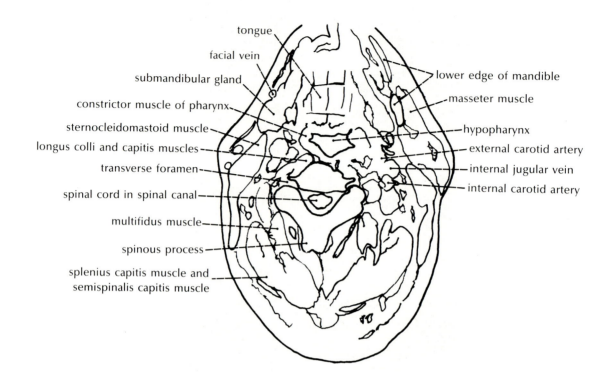

tongue
facial vein
submandibular gland
constrictor muscle of pharynx
sternocleidomastoid muscle
longus colli and capitis muscles
transverse foramen
spinal cord in spinal canal
multifidus muscle
spinous process
splenius capitis muscle and
semispinalis capitis muscle

lower edge of mandible
masseter muscle
hypopharynx
external carotid artery
internal jugular vein
internal carotid artery

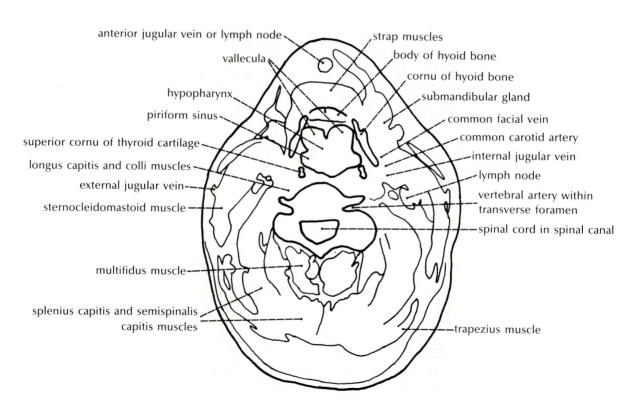

anterior jugular vein or lymph node
vallecula
hypopharynx
piriform sinus
superior cornu of thyroid cartilage
longus capitis and colli muscles
external jugular vein
sternocleidomastoid muscle
multifidus muscle
splenius capitis and semispinalis
capitis muscles

strap muscles
body of hyoid bone
cornu of hyoid bone
submandibular gland
common facial vein
common carotid artery
internal jugular vein
lymph node
vertebral artery within
transverse foramen
spinal cord in spinal canal
trapezius muscle

3/Neck, Computed Tomography

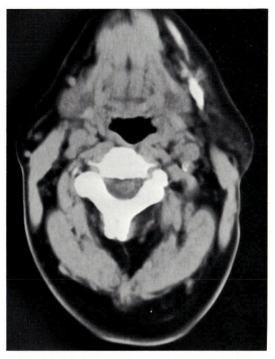

3-25 Section 4.

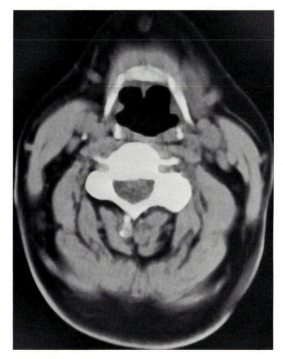

3-26 Section 5.

3/Neck, Computed Tomography

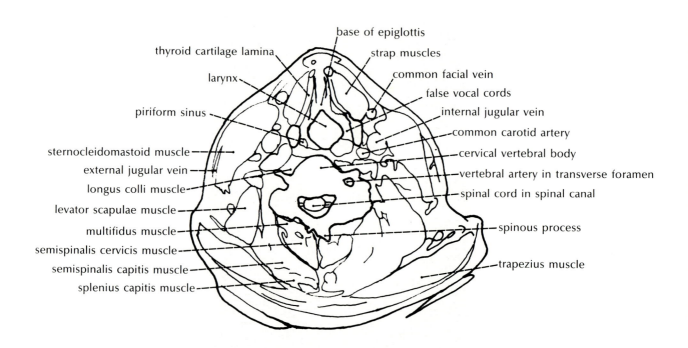

thyroid cartilage lamina
base of epiglottis
strap muscles
larynx
common facial vein
false vocal cords
piriform sinus
internal jugular vein
common carotid artery
sternocleidomastoid muscle
cervical vertebral body
external jugular vein
vertebral artery in transverse foramen
longus colli muscle
spinal cord in spinal canal
levator scapulae muscle
multifidus muscle
spinous process
semispinalis cervicis muscle
semispinalis capitis muscle
trapezius muscle
splenius capitis muscle

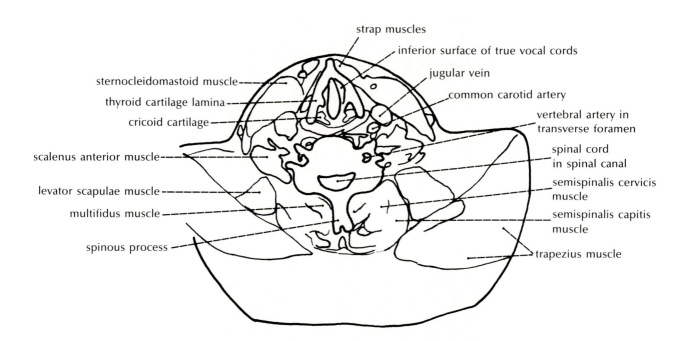

strap muscles
inferior surface of true vocal cords
sternocleidomastoid muscle
jugular vein
thyroid cartilage lamina
common carotid artery
cricoid cartilage
vertebral artery in transverse foramen
scalenus anterior muscle
spinal cord in spinal canal
levator scapulae muscle
semispinalis cervicis muscle
multifidus muscle
semispinalis capitis muscle
spinous process
trapezius muscle

3/Neck, Computed Tomography

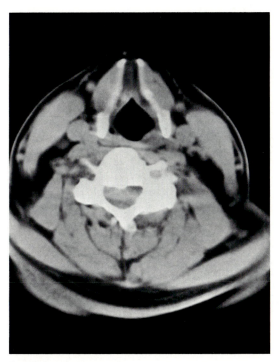

3-27 Section 6.

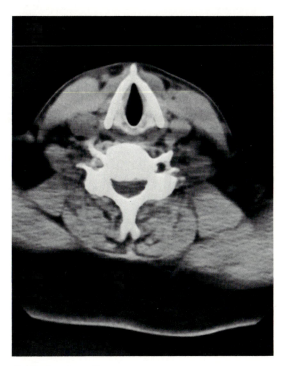

3-28 Section 7.

3/Thoracic Spine

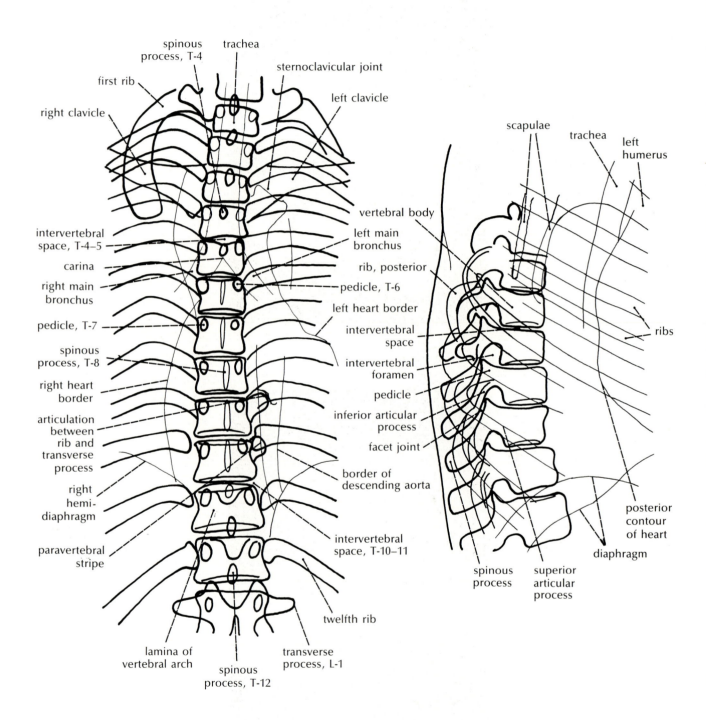

spinous process, T-4

trachea

sternoclavicular joint

first rib

left clavicle

right clavicle

vertebral body

left main bronchus

intervertebral space, T-4–5

rib, posterior

carina

pedicle, T-6

right main bronchus

left heart border

pedicle, T-7

intervertebral space

spinous process, T-8

intervertebral foramen

right heart border

pedicle

articulation between rib and transverse process

inferior articular process

facet joint

right hemi-diaphragm

border of descending aorta

paravertebral stripe

intervertebral space, T-10–11

twelfth rib

lamina of vertebral arch

transverse process, L-1

spinous process, T-12

scapulae

trachea

left humerus

ribs

posterior contour of heart

diaphragm

spinous process

superior articular process

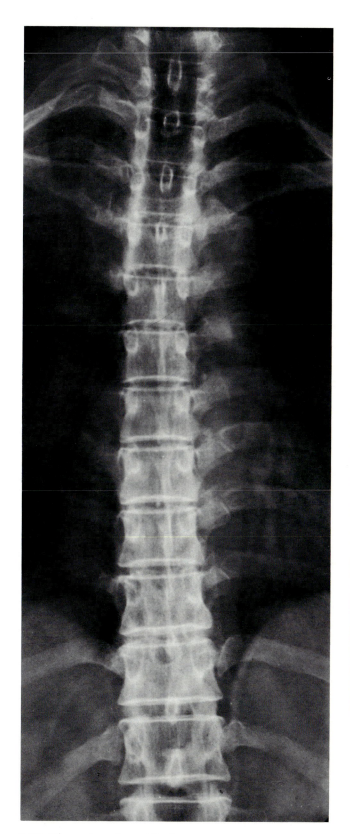

3-29 Thoracic spine, AP view.

3-30 Thoracic spine, lateral view.

3/Lumbar Spine

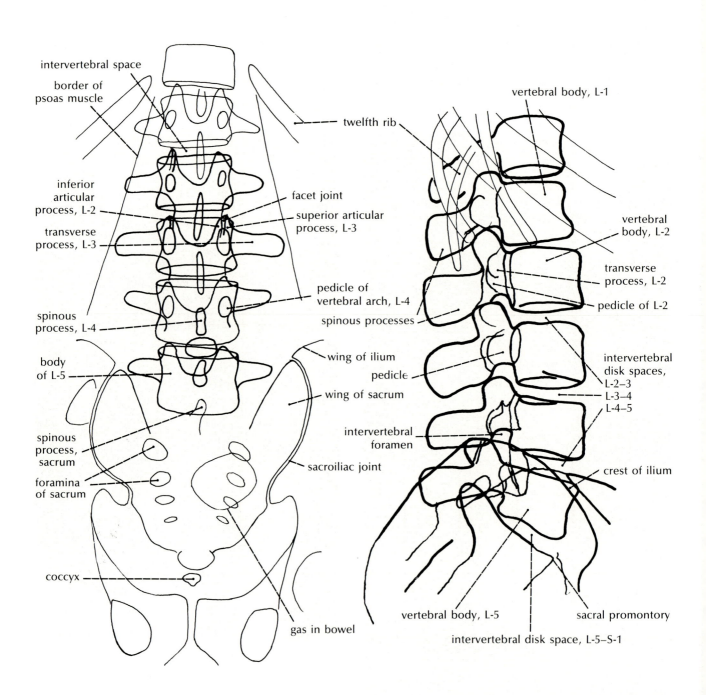

intervertebral space

border of
psoas muscle

twelfth rib

inferior
articular
process, L-2

facet joint

superior articular
process, L-3

transverse
process, L-3

spinous
process, L-4

pedicle of
vertebral arch, L-4

spinous processes

wing of ilium

body
of L-5

pedicle

wing of sacrum

spinous
process,
sacrum

intervertebral
foramen

foramina
of sacrum

sacroiliac joint

coccyx

gas in bowel

vertebral body, L-1

vertebral
body, L-2

transverse
process, L-2

pedicle of L-2

intervertebral
disk spaces,
L-2–3
L-3–4
L-4–5

crest of ilium

vertebral body, L-5

sacral promontory

intervertebral disk space, L-5–S-1

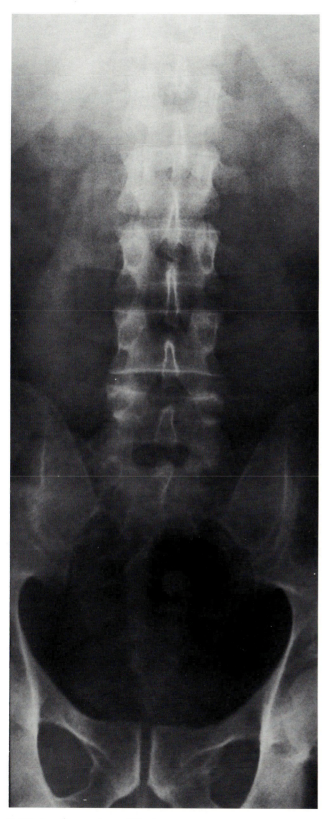

3-31 Lumbar spine, AP view.

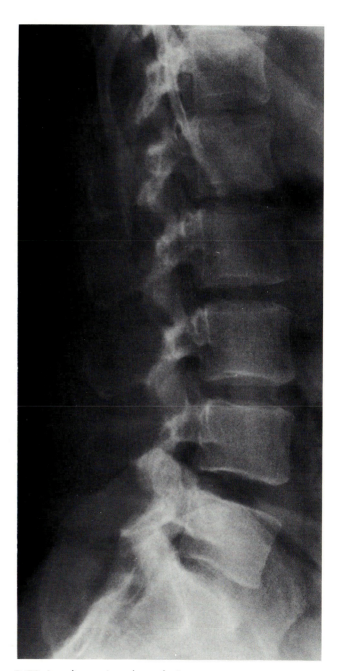

3-32 Lumbar spine, lateral view.

3/Lumbar Spine; Sacroiliac Joint

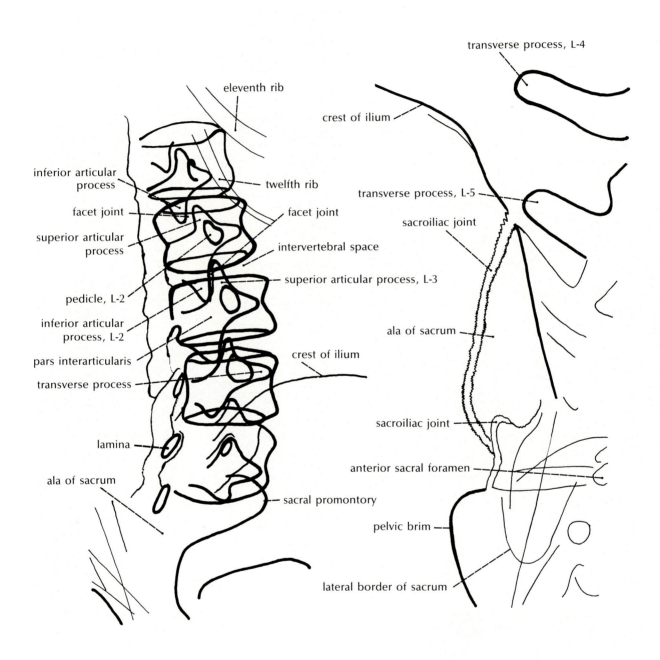

eleventh rib

crest of ilium

transverse process, L-4

inferior articular process

twelfth rib

facet joint

facet joint

transverse process, L-5

superior articular process

intervertebral space

sacroiliac joint

superior articular process, L-3

pedicle, L-2

ala of sacrum

inferior articular process, L-2

pars interarticularis

crest of ilium

transverse process

sacroiliac joint

anterior sacral foramen

lamina

pelvic brim

ala of sacrum

sacral promontory

lateral border of sacrum

3/Lumbar Spine; Sacroiliac Joint

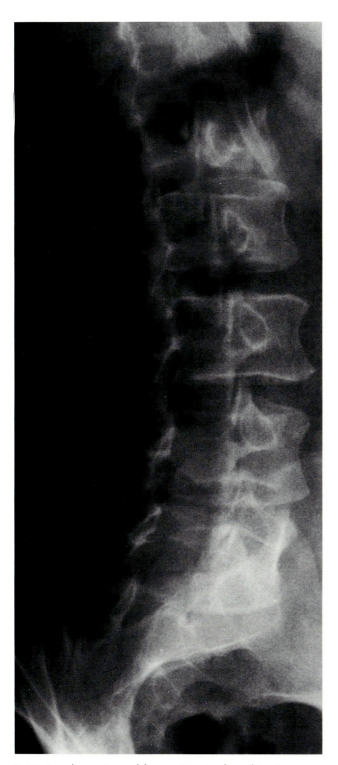

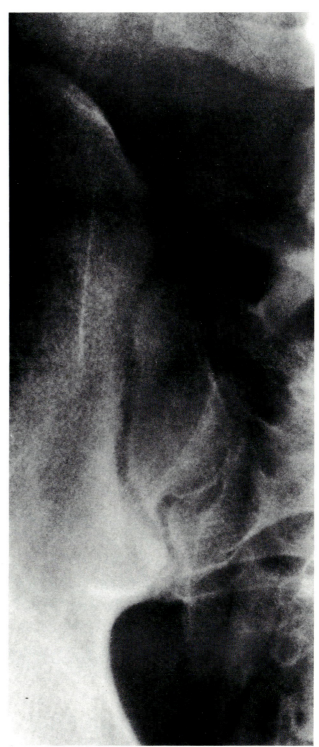

3-33 Lumbar spine, oblique view (reduced in size compared with Fig. 3-34).

3-34 Sacroiliac joint, oblique view.

3/Sacrum and Coccyx

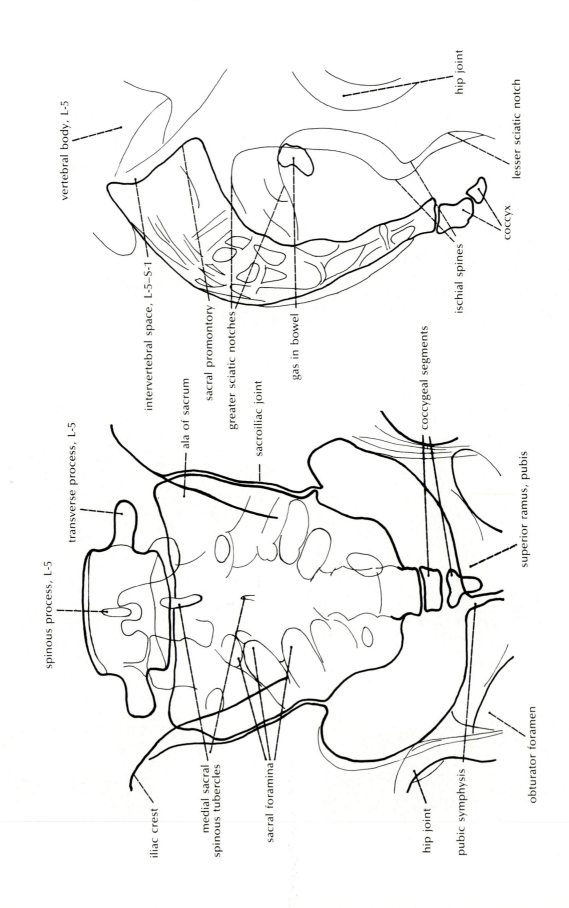

3/Sacrum and Coccyx

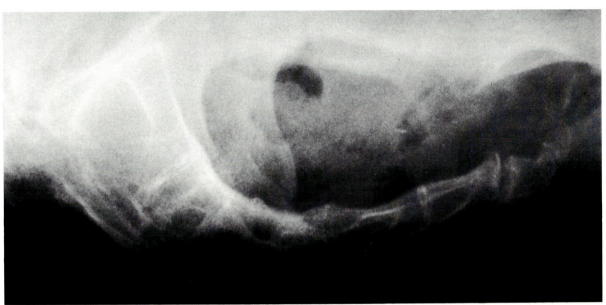

3-36 Sacrum and coccyx, lateral view.

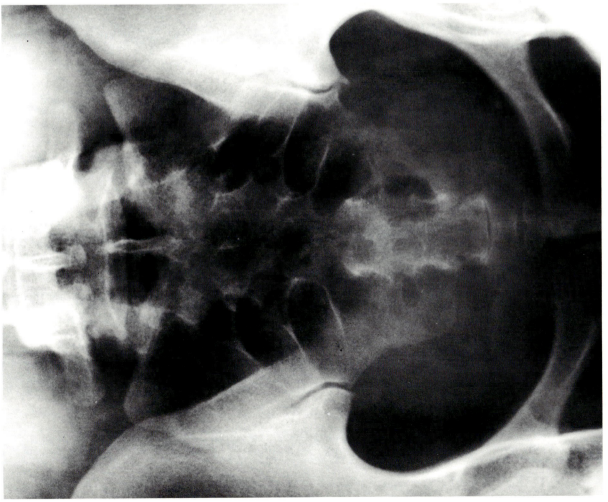

3-35 Sacrum and coccyx, AP view.

3/Thyroid

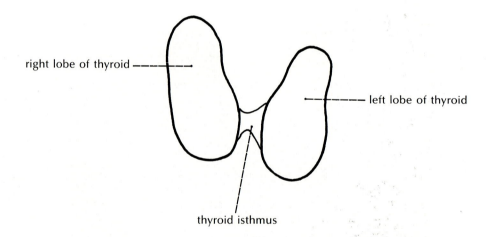

right lobe of thyroid

left lobe of thyroid

thyroid isthmus

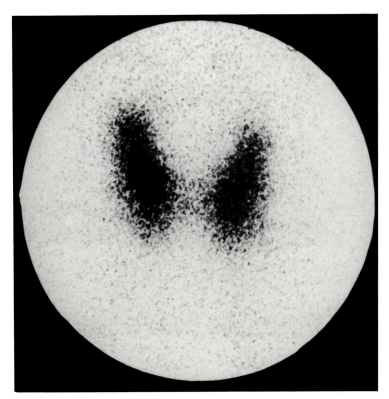

3-37 Thyroid gland, nuclear scan.

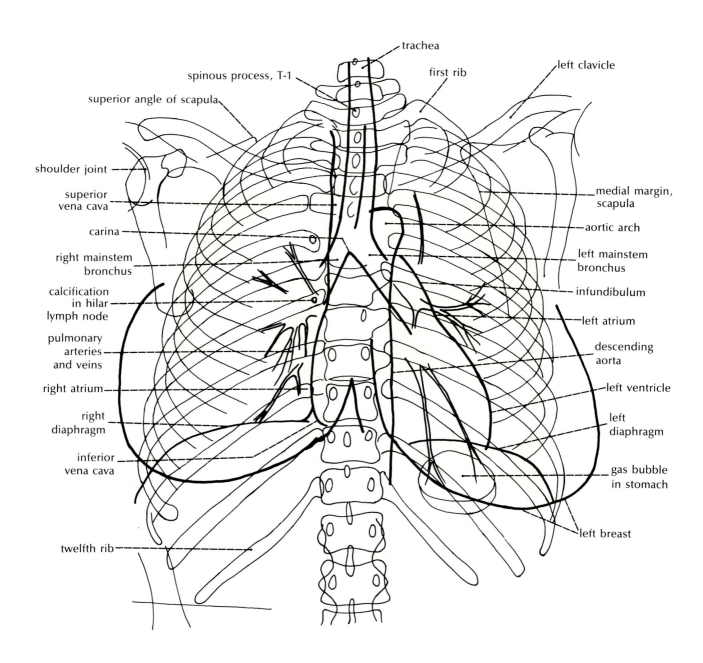

trachea

spinous process, T-1

first rib

left clavicle

superior angle of scapula

shoulder joint

superior vena cava

medial margin, scapula

carina

aortic arch

right mainstem bronchus

left mainstem bronchus

calcification in hilar lymph node

infundibulum

left atrium

pulmonary arteries and veins

descending aorta

right atrium

left ventricle

right diaphragm

left diaphragm

inferior vena cava

gas bubble in stomach

left breast

twelfth rib

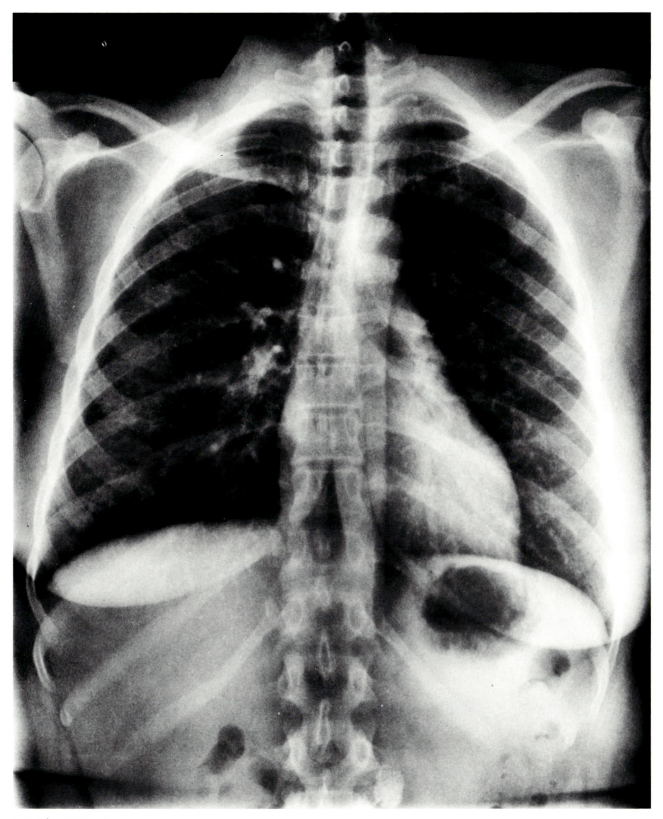

4-1 Chest, PA view.

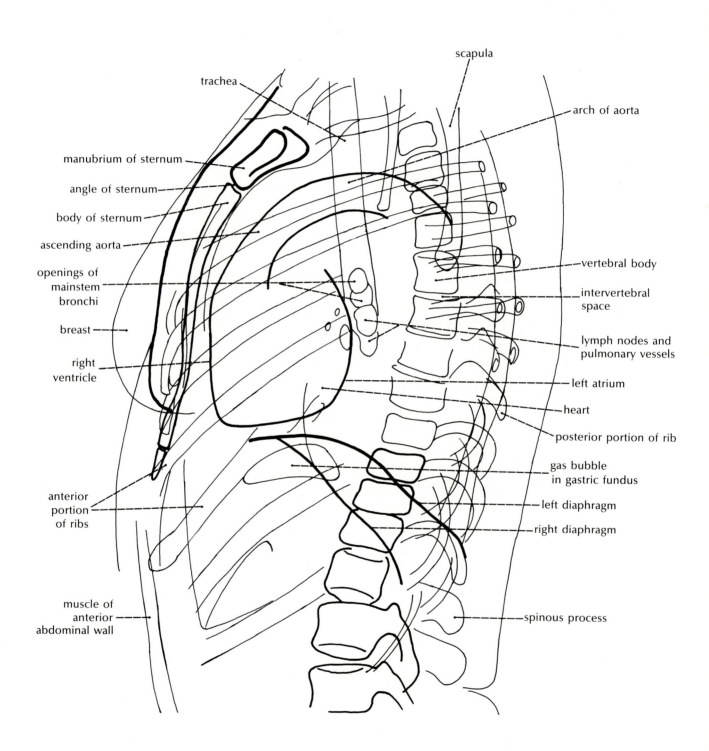

scapula

trachea

arch of aorta

manubrium of sternum

angle of sternum

body of sternum

ascending aorta

vertebral body

openings of
mainstem
bronchi

intervertebral
space

breast

lymph nodes and
pulmonary vessels

right
ventricle

left atrium

heart

posterior portion of rib

gas bubble
in gastric fundus

anterior
portion
of ribs

left diaphragm

right diaphragm

muscle of
anterior
abdominal wall

spinous process

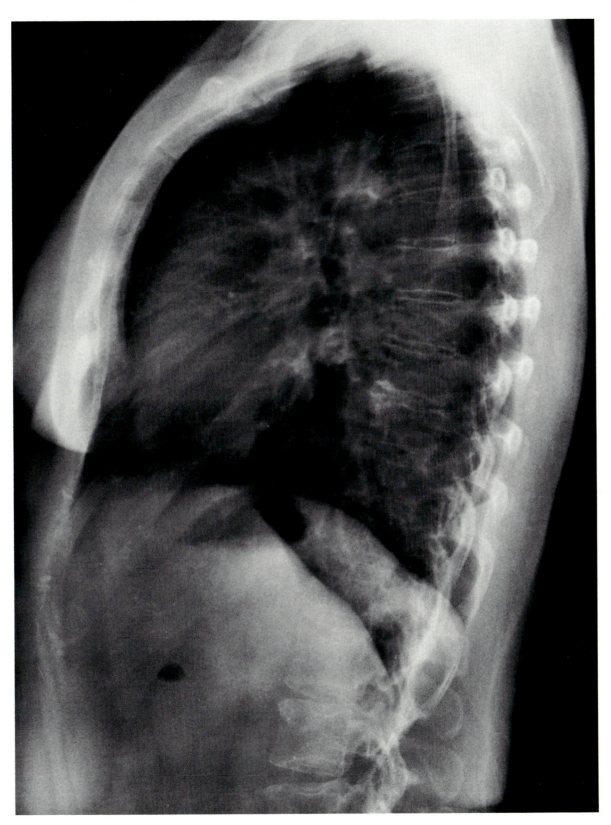

4-2 Chest, lateral view.

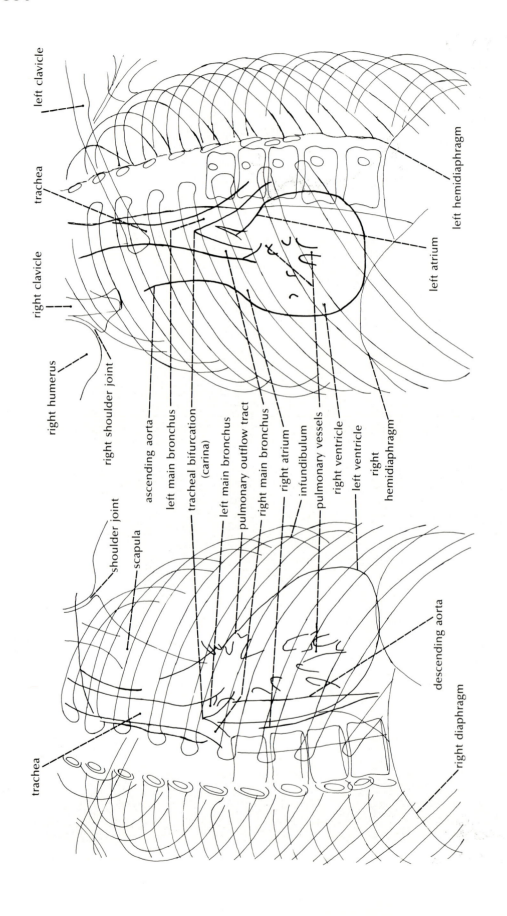

left clavicle

trachea

right clavicle

right humerus

right shoulder joint

left hemidiaphragm

left atrium

shoulder joint

scapula

ascending aorta

left main bronchus

tracheal bifurcation (carina)

left main bronchus

pulmonary outflow tract

right main bronchus

right atrium

infundibulum

pulmonary vessels

right ventricle

left ventricle

right hemidiaphragm

trachea

descending aorta

right diaphragm

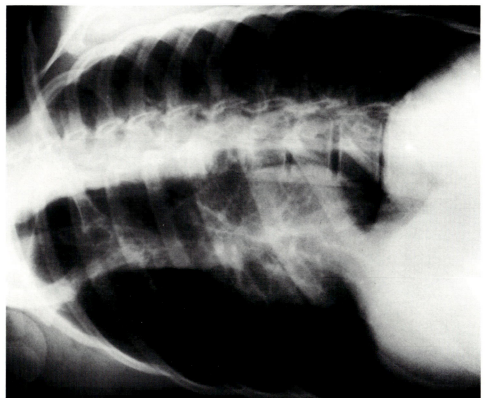

4-4 Chest, left anterior oblique (LAO) view.

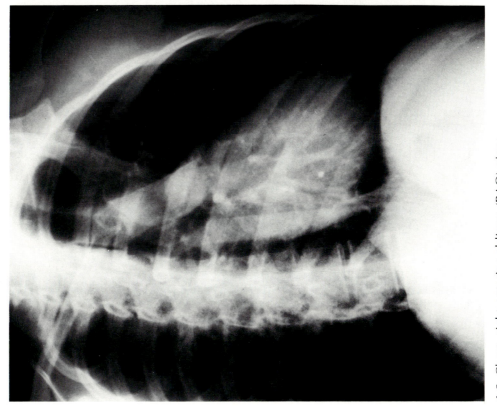

4-3 Chest, right anterior oblique (RAO) view.

4/Angiocardiography

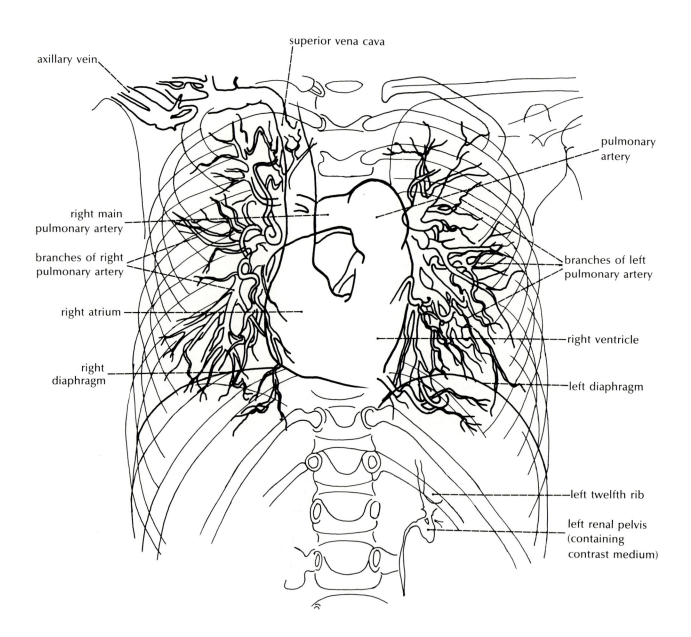

axillary vein

superior vena cava

pulmonary artery

right main pulmonary artery

branches of right pulmonary artery

right atrium

right diaphragm

branches of left pulmonary artery

right ventricle

left diaphragm

left twelfth rib

left renal pelvis (containing contrast medium)

4/Angiocardiography

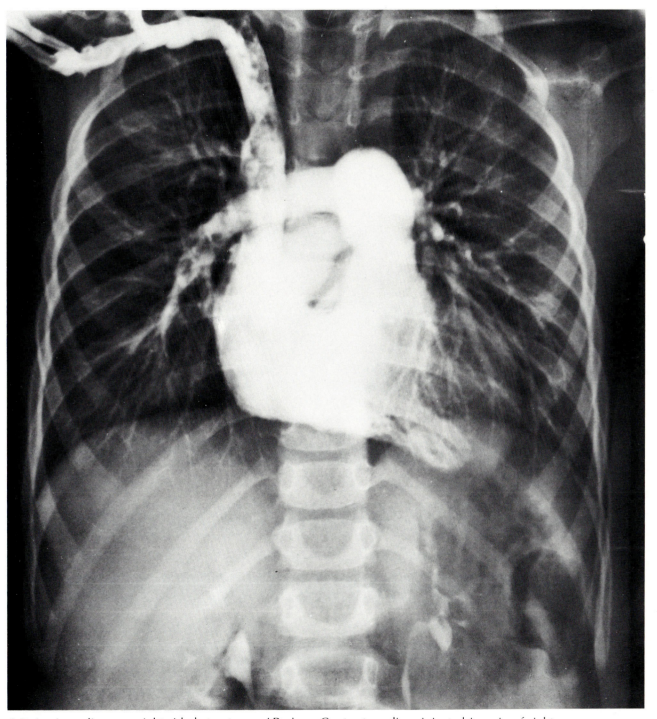

4-5 Angiocardiogram, right-sided structures. AP view. Contrast medium injected in vein of right arm.

4/Angiocardiography

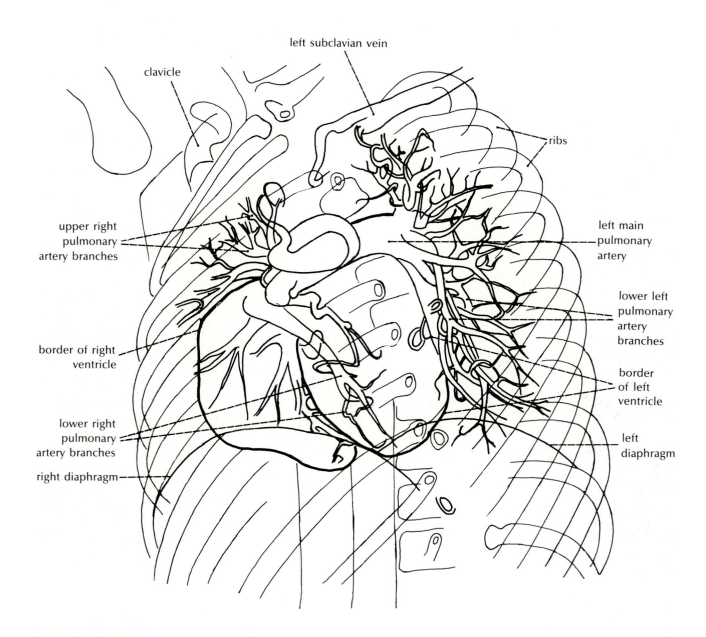

left subclavian vein

clavicle

ribs

upper right
pulmonary
artery branches

left main
pulmonary
artery

lower left
pulmonary
artery
branches

border of right
ventricle

border
of left
ventricle

lower right
pulmonary
artery branches

left
diaphragm

right diaphragm

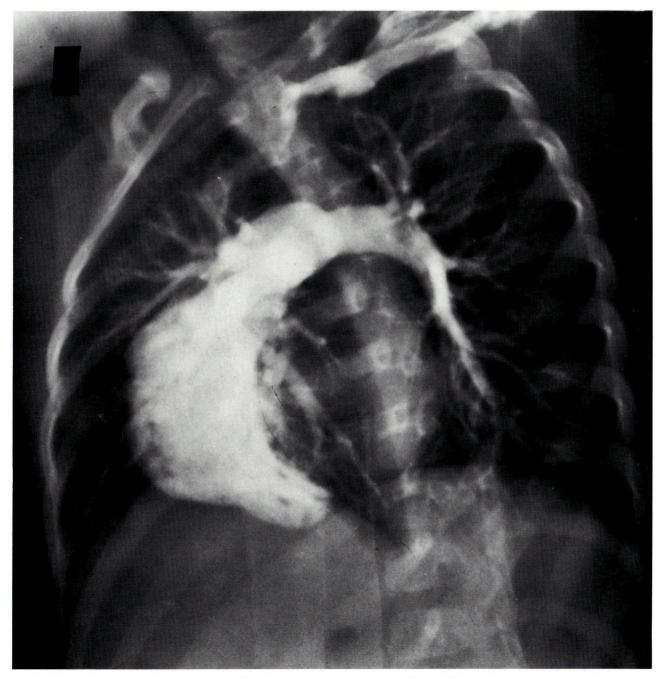

4-6 Angiocardiogram, right ventricle and pulmonary artery and branches. Oblique view. Contrast medium injected in vein of right arm.

4/Angiocardiography

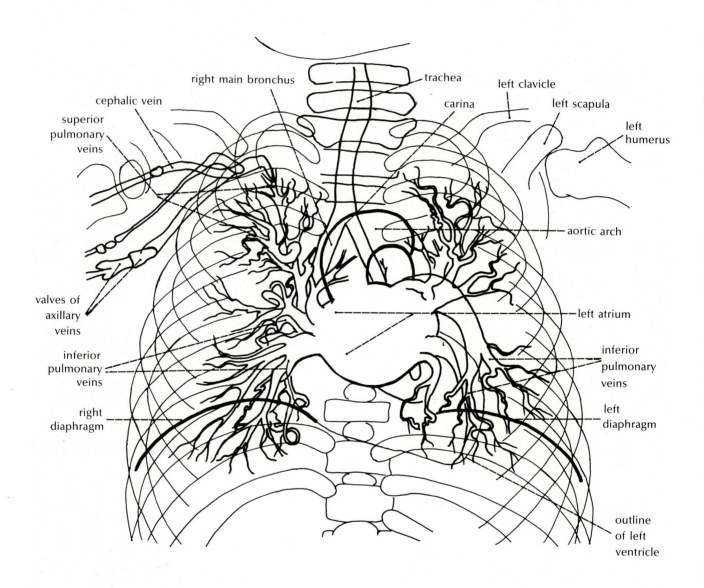

right main bronchus

trachea

cephalic vein

carina

left clavicle

superior
pulmonary
veins

left scapula

left
humerus

aortic arch

valves of
axillary
veins

left atrium

inferior
pulmonary
veins

inferior
pulmonary
veins

right
diaphragm

left
diaphragm

outline
of left
ventricle

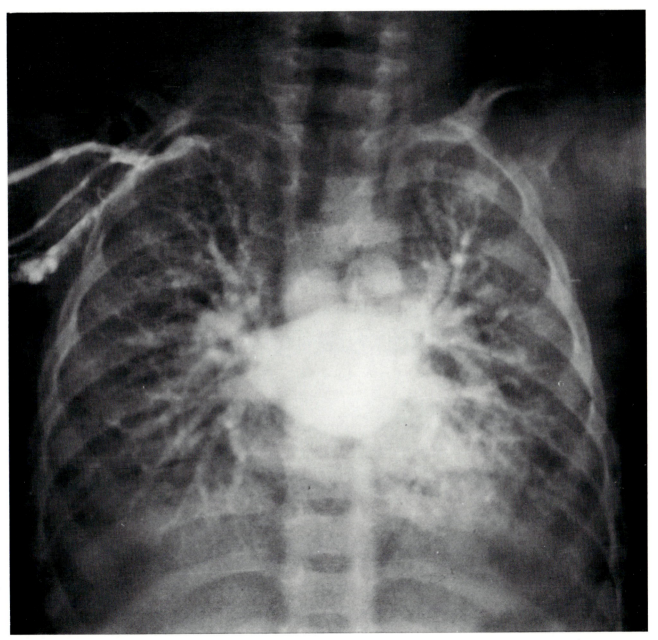

4-7 Angiocardiogram, left pulmonary veins and left atrium. AP view of a young child. Contrast medium injected in vein of right arm.

4/Angiocardiography

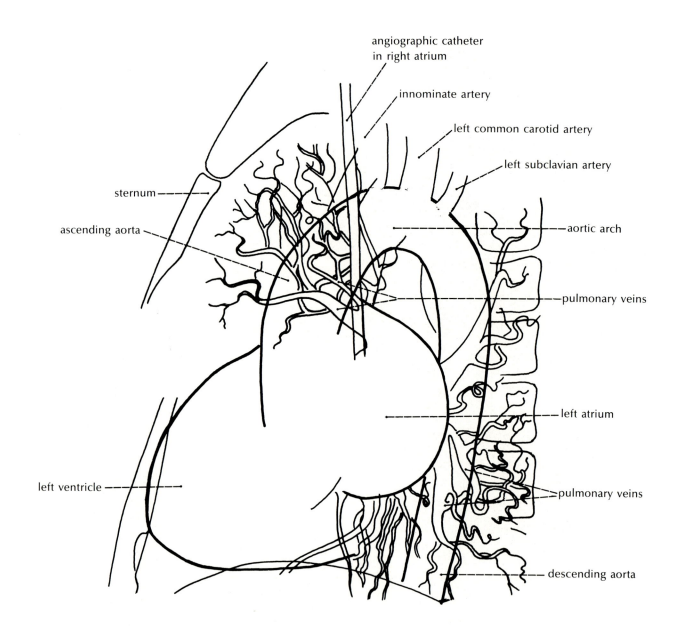

angiographic catheter
in right atrium

innominate artery

left common carotid artery

left subclavian artery

sternum

aortic arch

ascending aorta

pulmonary veins

left atrium

left ventricle

pulmonary veins

descending aorta

4/Angiocardiography

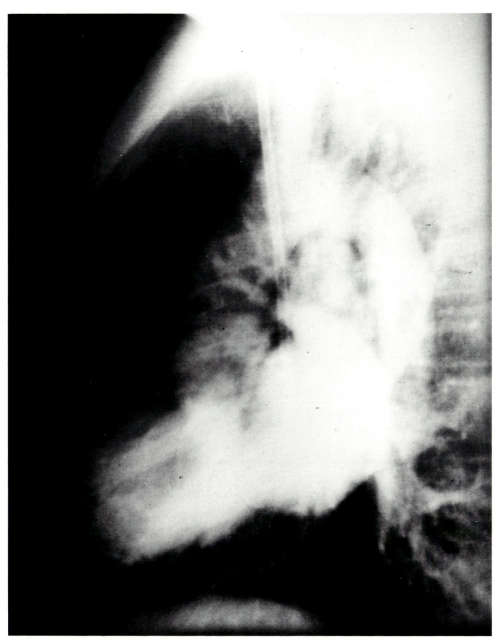

4-8 Angiocardiogram, left atrium, left ventricle and aorta and branches from the arch. Lateral view. Contrast medium injected from catheter in right atrium.

4/Angiocardiography

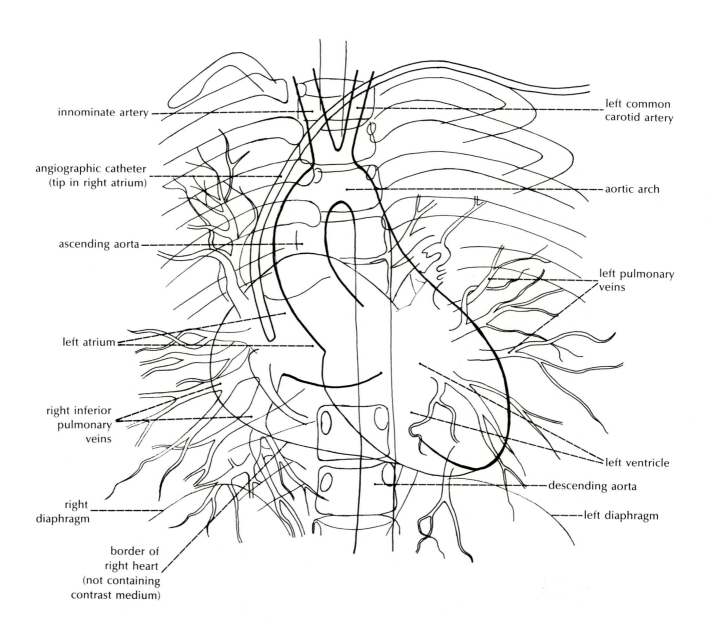

innominate artery

left common
carotid artery

angiographic catheter
(tip in right atrium)

aortic arch

ascending aorta

left pulmonary
veins

left atrium

right inferior
pulmonary
veins

left ventricle

descending aorta

right
diaphragm

left diaphragm

border of
right heart
(not containing
contrast medium)

4/Angiocardiography

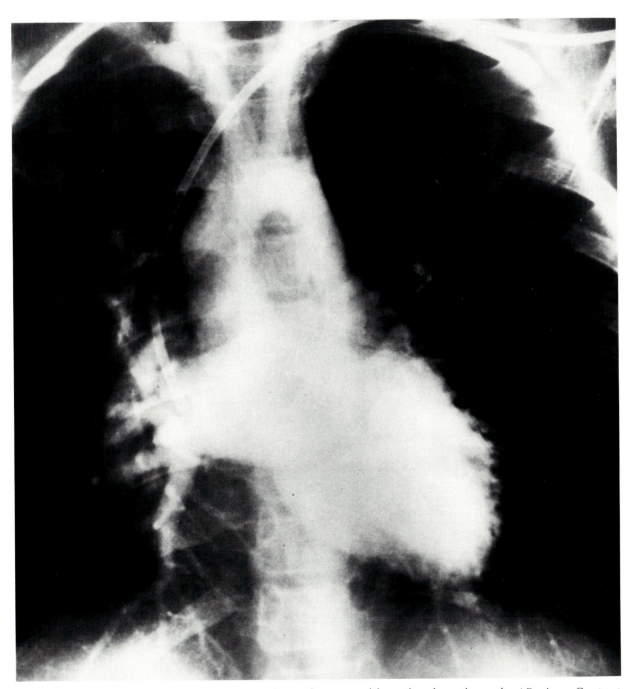

4-9 Angiocardiogram, left atrium, left ventricle, and aorta and branches from the arch. AP view. Contrast medium injected from catheter in right atrium.

4/Coronary Arteriography

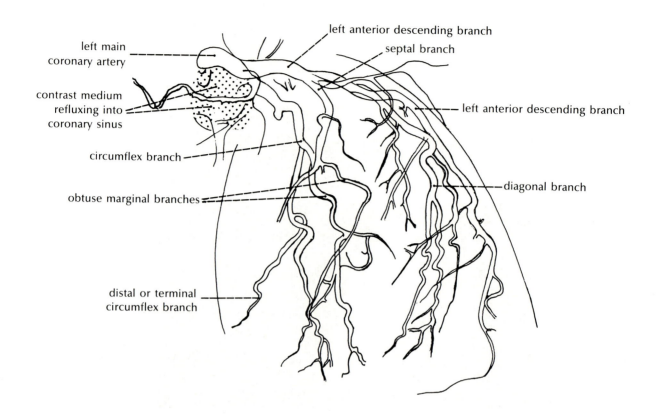

left main coronary artery

contrast medium refluxing into coronary sinus

circumflex branch

obtuse marginal branches

distal or terminal circumflex branch

left anterior descending branch

septal branch

left anterior descending branch

diagonal branch

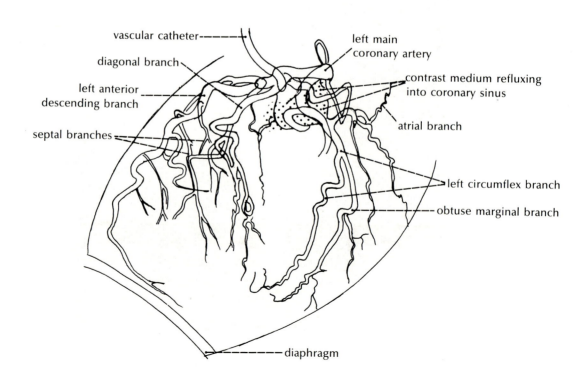

vascular catheter

diagonal branch

left anterior descending branch

septal branches

left main coronary artery

contrast medium refluxing into coronary sinus

atrial branch

left circumflex branch

obtuse marginal branch

diaphragm

4/Coronary Arteriography

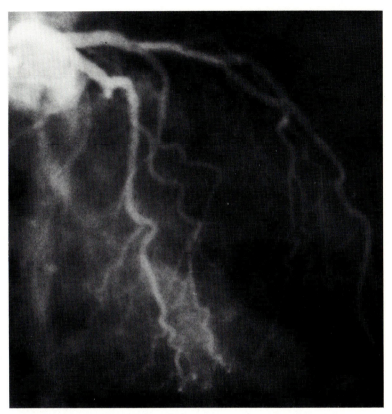

4-10 Coronary arteriogram, left coronary artery, RAO view.

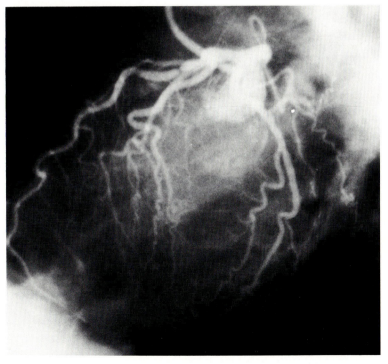

4-11 Coronary arteriogram, left coronary artery, lateral view.

4/Coronary Arteriography

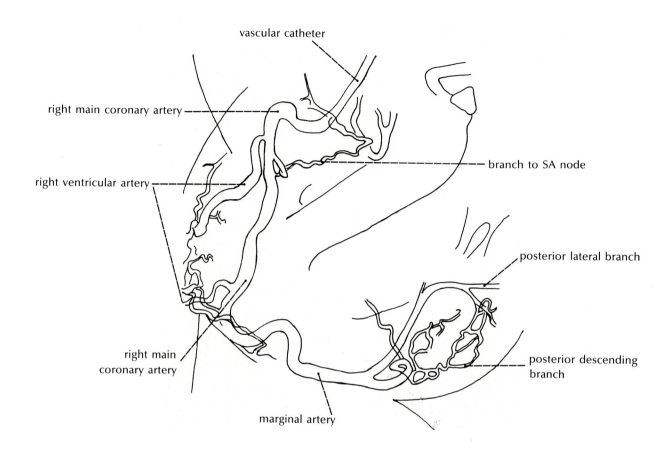

vascular catheter

right main coronary artery

branch to SA node

right ventricular artery

posterior lateral branch

right main
coronary artery

posterior descending
branch

marginal artery

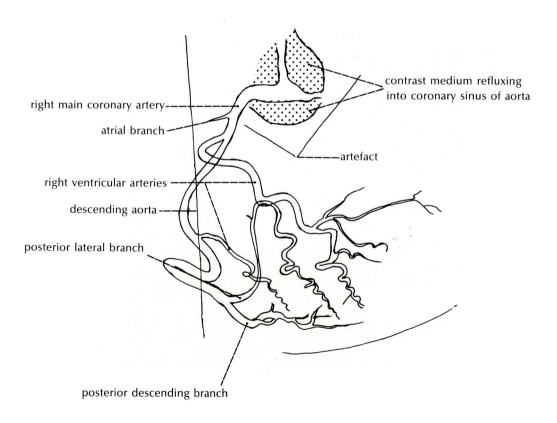

right main coronary artery

contrast medium refluxing
into coronary sinus of aorta

atrial branch

artefact

right ventricular arteries

descending aorta

posterior lateral branch

posterior descending branch

4/Coronary Arteriography

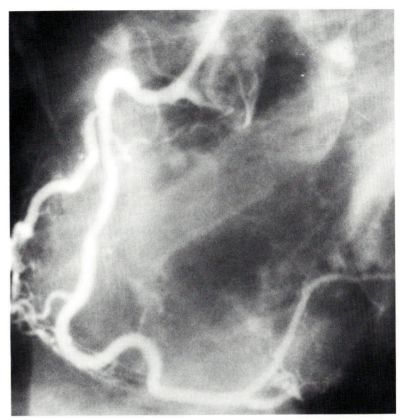

4-12 Coronary arteriogram, right coronary artery, lateral view.

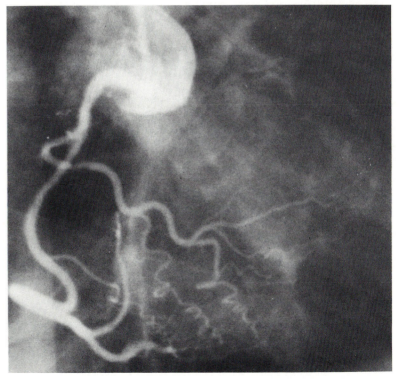

4-13 Coronary arteriogram, right coronary artery, RAO view.

4/Aortic Arch, Angiography

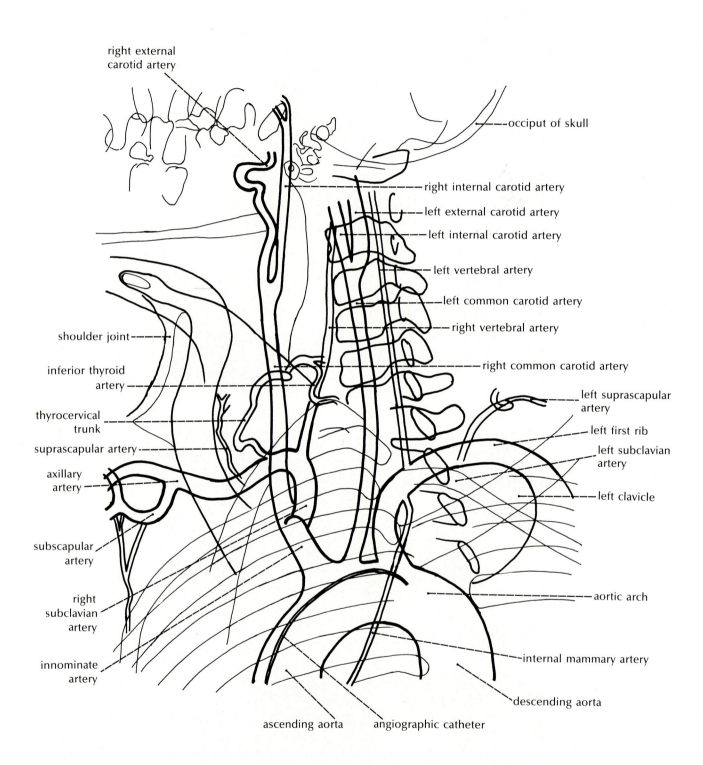

right external
carotid artery

occiput of skull

right internal carotid artery

left external carotid artery

left internal carotid artery

left vertebral artery

left common carotid artery

right vertebral artery

shoulder joint

right common carotid artery

inferior thyroid
artery

left suprascapular
artery

thyrocervical
trunk

left first rib

left subclavian
artery

suprascapular artery

axillary
artery

left clavicle

subscapular
artery

right
subclavian
artery

aortic arch

internal mammary artery

innominate
artery

descending aorta

ascending aorta   angiographic catheter

4/Aortic Arch, Angiography

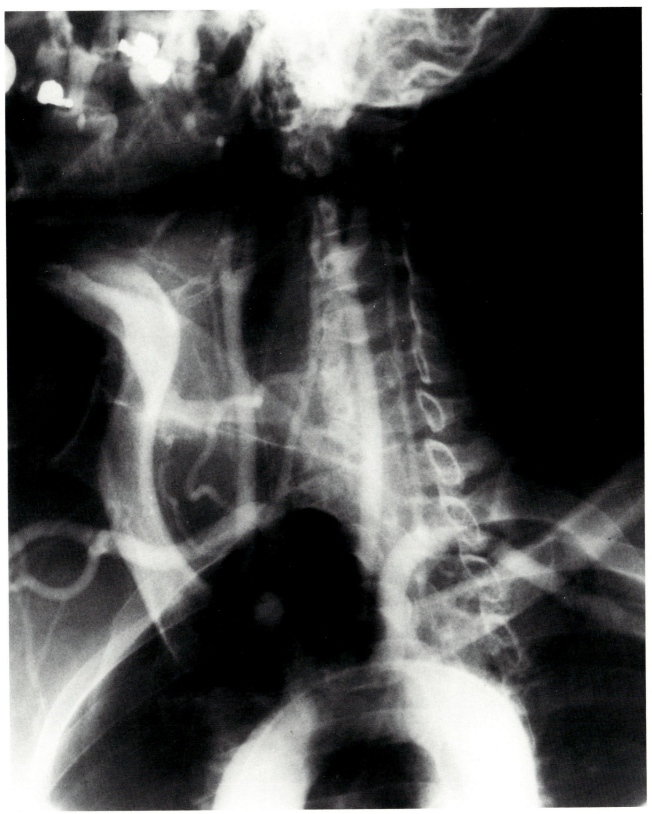

4-14 Angiogram, aortic arch and its branches. Contrast medium injected from catheter in the proximal aortic arch.

4/Bronchography

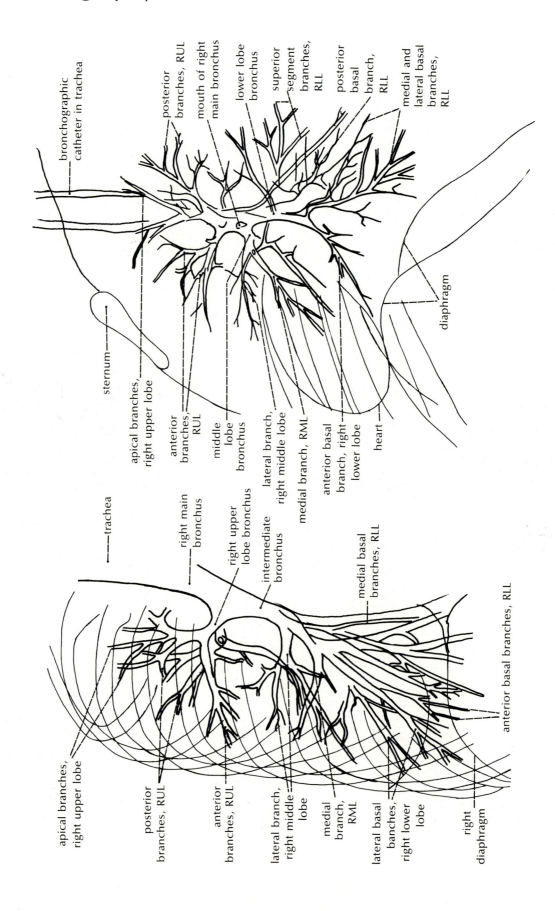

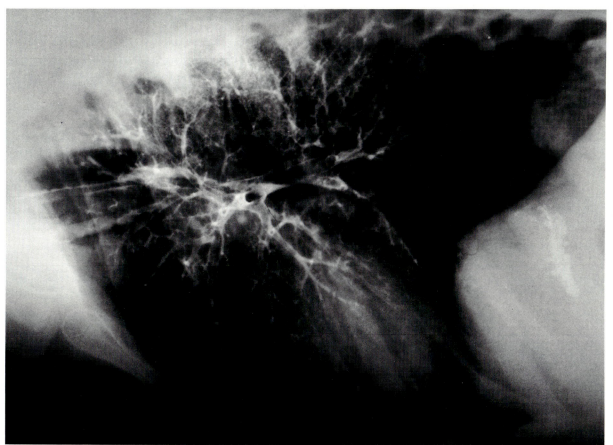

4-16 Right bronchogram, lateral view. Contrast medium injected in trachea.

4-15 Right bronchogram, AP view. Contrast medium injected in trachea.

4/Bronchography

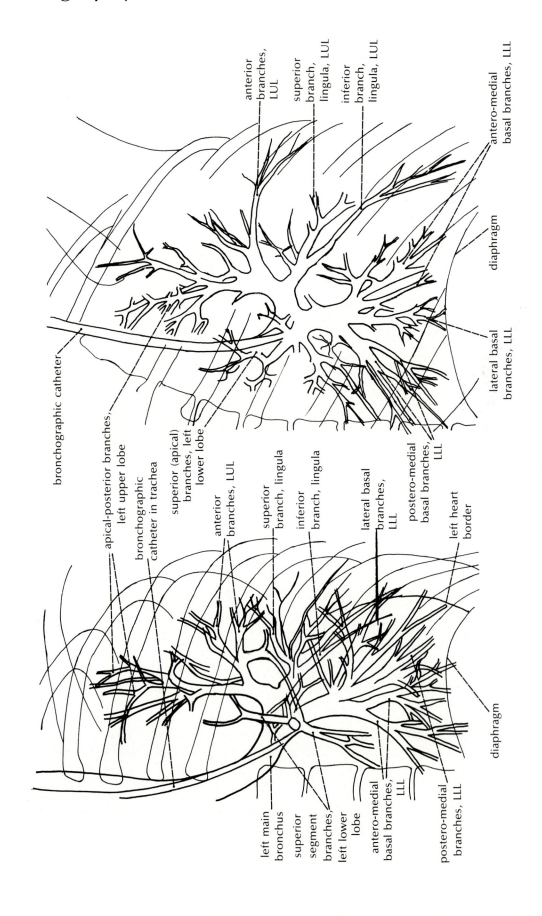

4/Bronchography

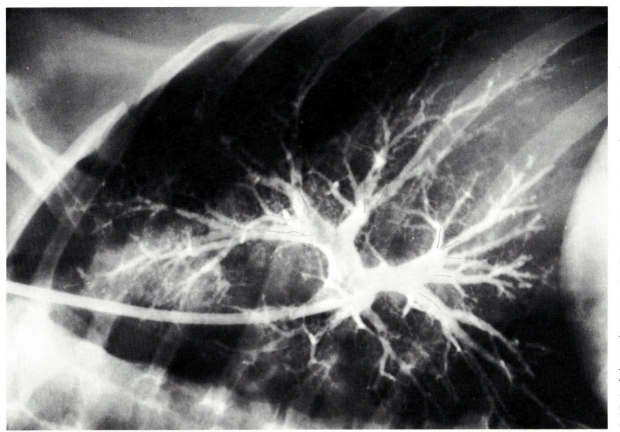

4-18 Left bronchogram, RAO view. Contrast medium injected in trachea.

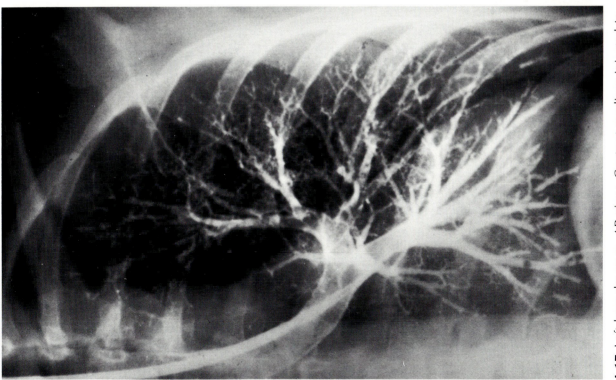

4-17 Left bronchogram, AP view. Contrast medium injected in trachea.

4/Thoracic Esophagus

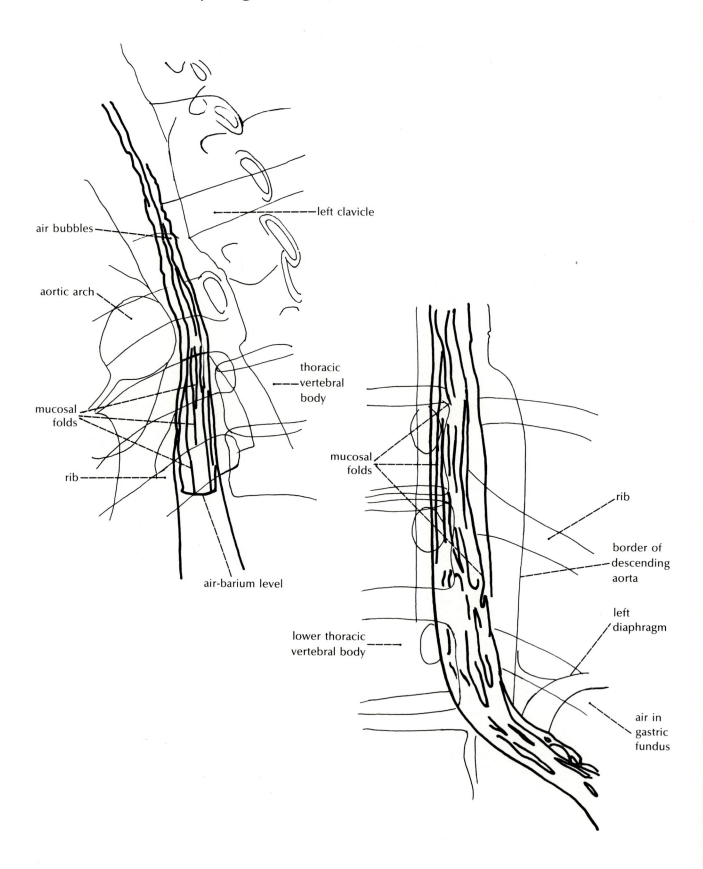

left clavicle

air bubbles

aortic arch

thoracic vertebral body

mucosal folds

rib

air-barium level

mucosal folds

rib

border of descending aorta

left diaphragm

lower thoracic vertebral body

air in gastric fundus

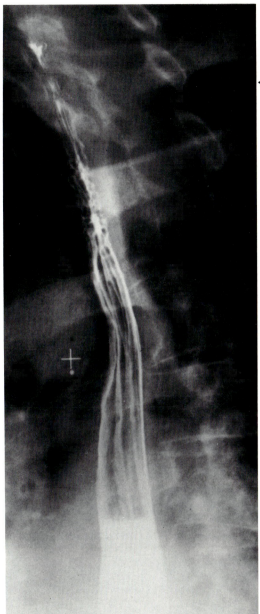

◄Upper esophagus

Lower esophagus►

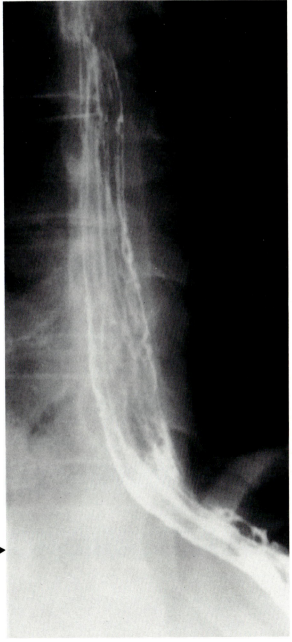

4-19 Thoracic esophagus, LAO view.

4/Breast, Female

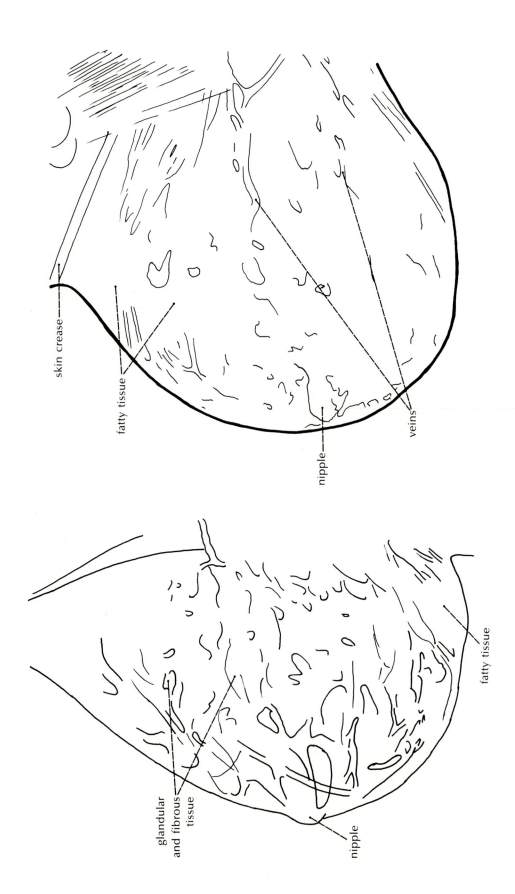

skin crease

fatty tissue

nipple

veins

glandular and fibrous tissue

nipple

fatty tissue

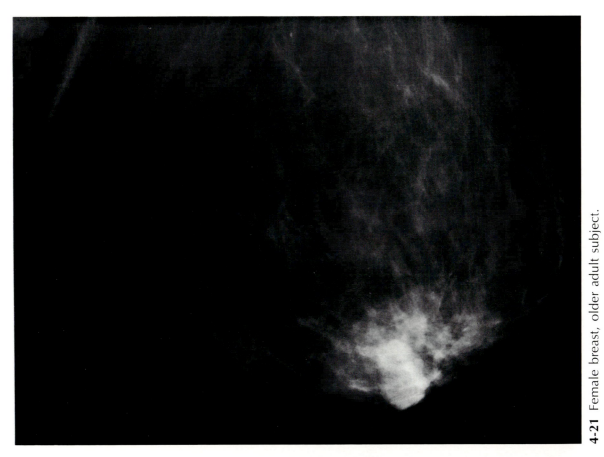

4-21 Female breast, older adult subject.

4-20 Female breast, younger adult subject.

4/Chest, Computed Tomography

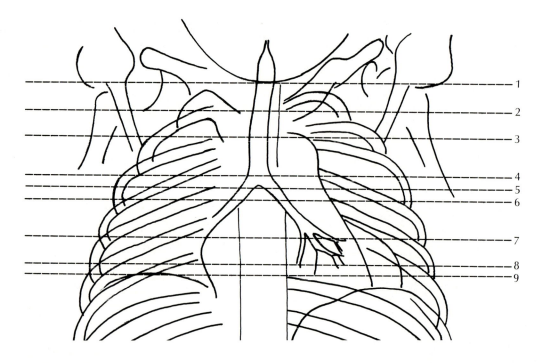

CT slices of the chest shown in Figs. 4-23 to 4-35. Figs. 4-23, 24, 25, 27, 28, 30, 32, and 34 were made with the window settings of the CT machine to enhance details of the mediastinum and the soft tissues of the chest wall. Figs. 4-26, 29, 31, 33, and 35 were made with the window settings to enhance the details of the lungs.

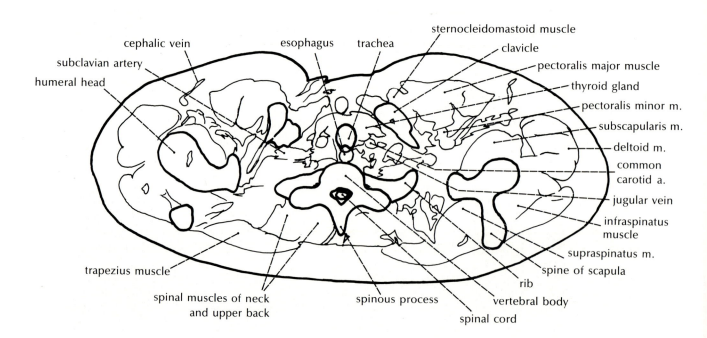

4/Chest, Computed Tomography

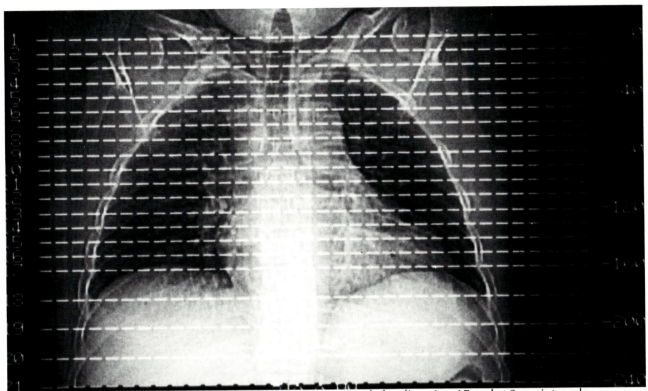

4-22 Slices made by CT examination of the chest, at 1-cm intervals for slices 1 to 17 and at 2-cm intervals for slices 17 to 21. See drawing (opposite) for those slices out of these 21 that are shown in Figs. 4-23 through 4-35.

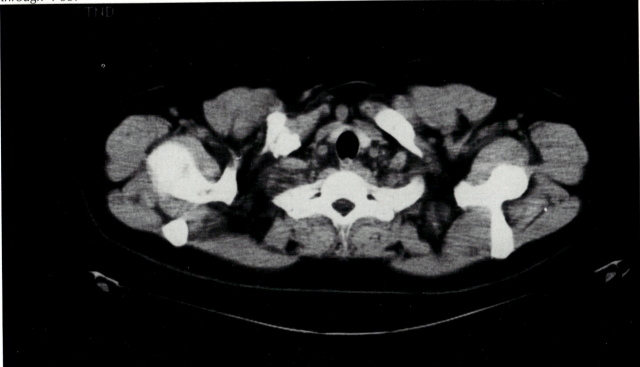

4-23 Section 1 [slice 1 on drawing (p. 118) accompanying Fig. 4-22].

4/Chest, Computed Tomography

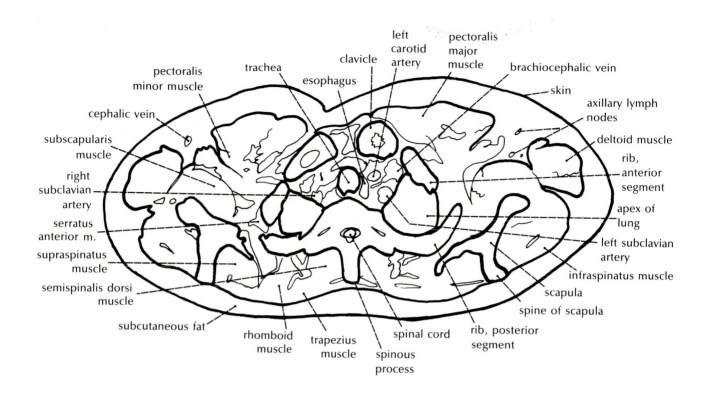

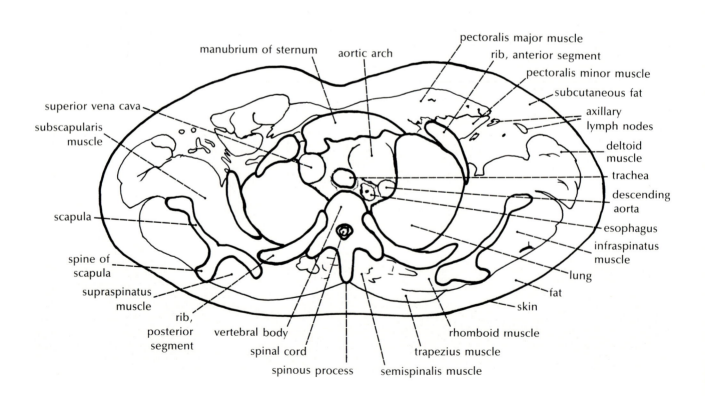

4/Chest, Computed Tomography

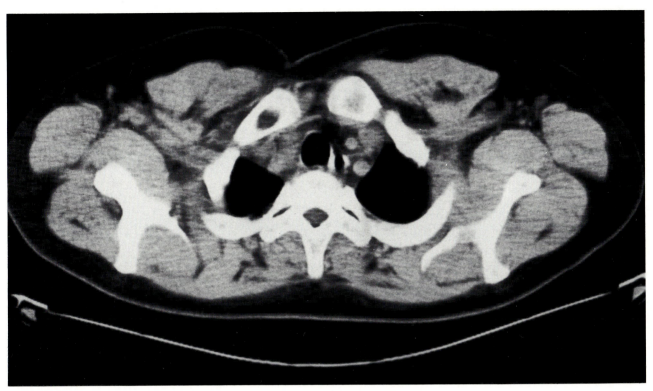

4-24 Section 2 [slice 2 on drawing (p. 118) accompanying Fig. 4-22].

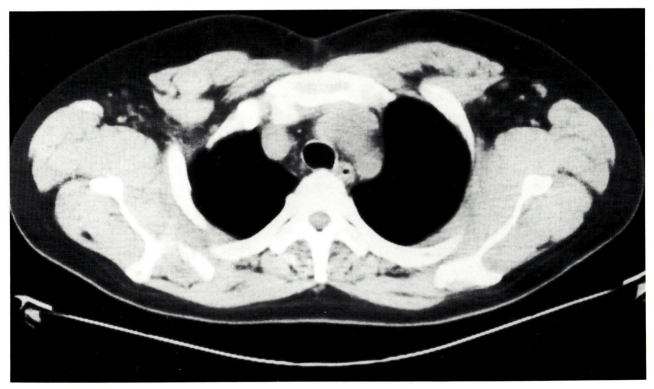

4-25 Section 3 [slice 3 on drawing (p. 118) accompanying Fig. 4-22].

4/Chest, Computed Tomography

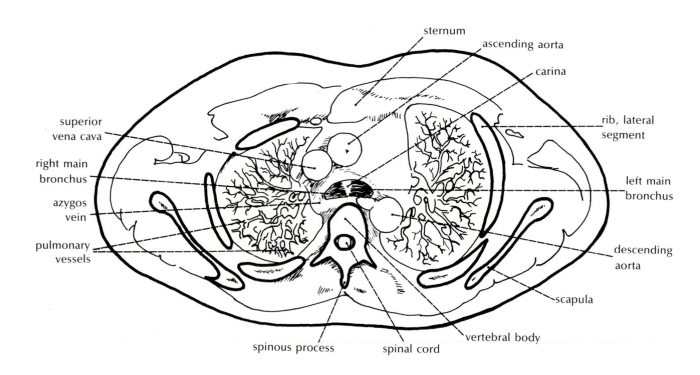

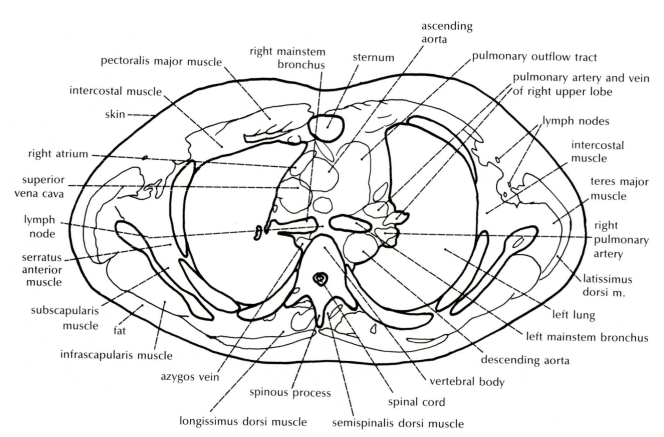

4/Chest, Computed Tomography

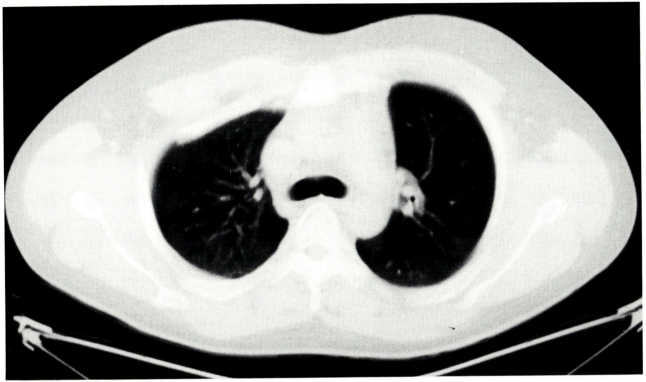

4-26 Section 4 [slice 4 on drawing (p. 118) accompanying Fig. 4-22].

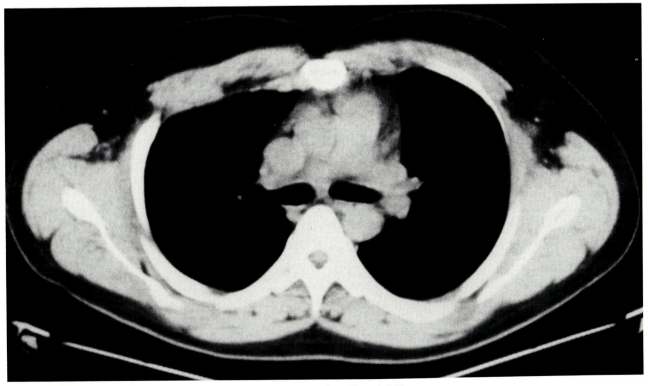

4-27 Section 5 [slice 5 on drawing (p. 118) accompanying Fig. 4-22].

4/Chest, Computed Tomography

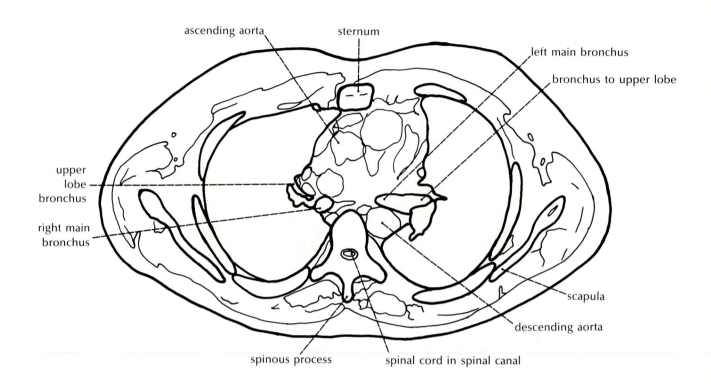

ascending aorta   sternum   left main bronchus

bronchus to upper lobe

upper lobe bronchus

right main bronchus

scapula

descending aorta

spinous process   spinal cord in spinal canal

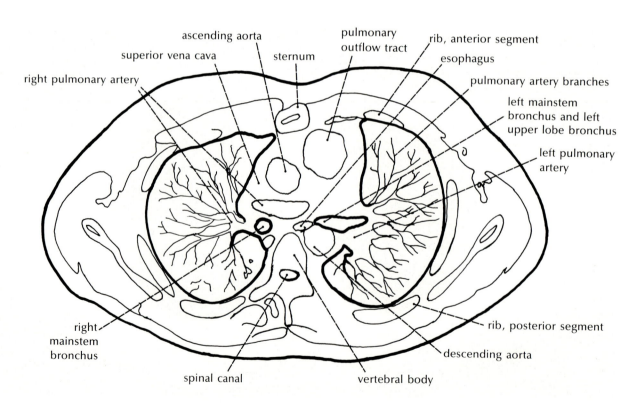

ascending aorta   pulmonary outflow tract   rib, anterior segment

superior vena cava   sternum   esophagus

right pulmonary artery   pulmonary artery branches

left mainstem bronchus and left upper lobe bronchus

left pulmonary artery

right mainstem bronchus

rib, posterior segment

descending aorta

spinal canal   vertebral body

4/Chest, Computed Tomography

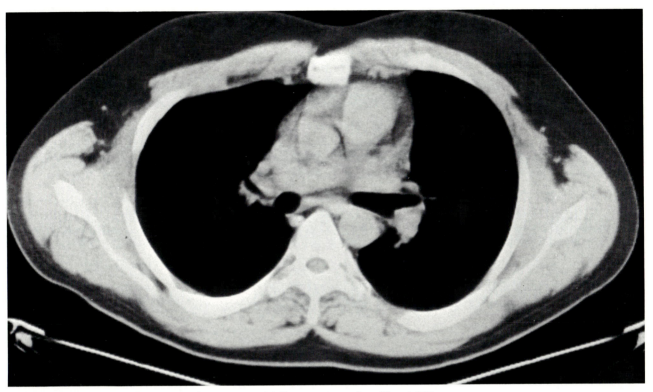

4-28 Section 6 [from slice 6 on drawing (p. 118) accompanying Fig. 4-22].

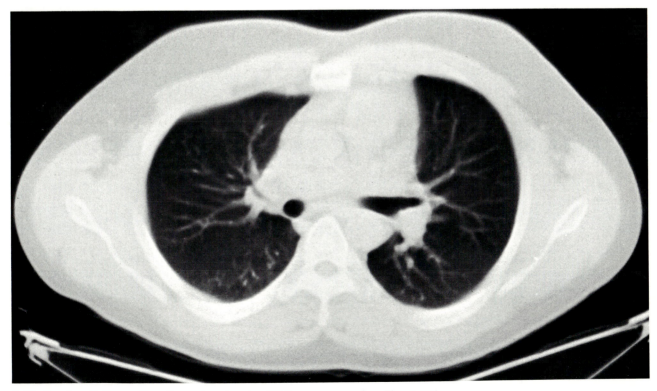

4-29 Section 7 [from slice 6 on drawing (p. 118) accompanying Fig. 4-22].

4/Chest, Computed Tomography

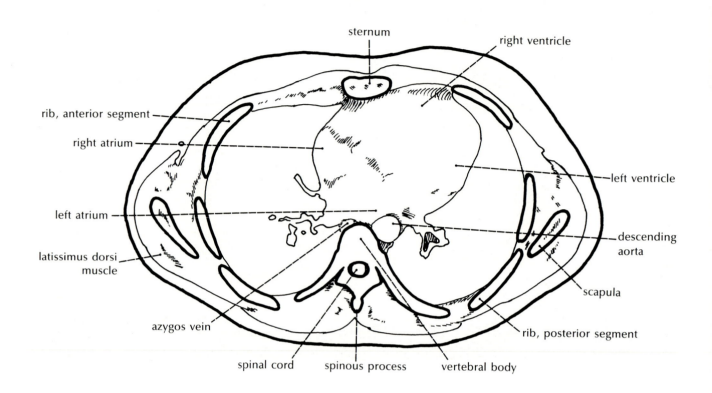

sternum

right ventricle

rib, anterior segment

right atrium

left ventricle

left atrium

descending aorta

latissimus dorsi muscle

scapula

azygos vein

rib, posterior segment

spinal cord

spinous process

vertebral body

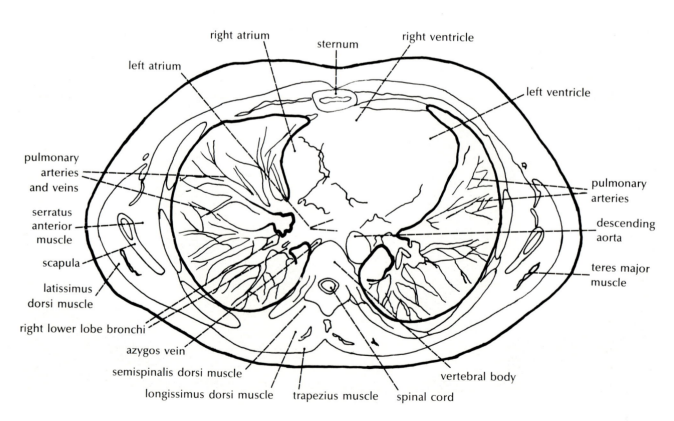

right atrium

sternum

right ventricle

left atrium

left ventricle

pulmonary arteries and veins

pulmonary arteries

serratus anterior muscle

descending aorta

scapula

teres major muscle

latissimus dorsi muscle

right lower lobe bronchi

azygos vein

semispinalis dorsi muscle

vertebral body

longissimus dorsi muscle

trapezius muscle

spinal cord

4/Chest, Computed Tomography

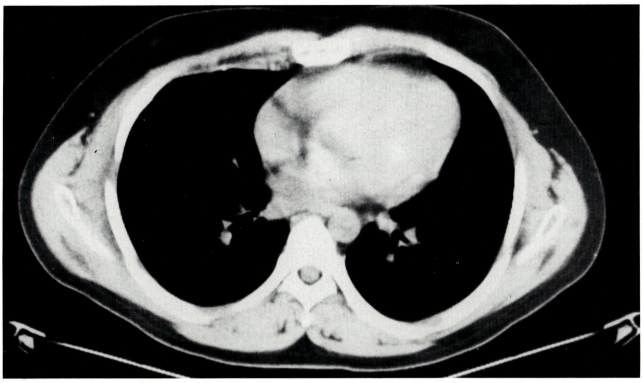

4-30 Section 8 [from slice 7 on drawing (p. 118) accompanying Fig. 4-22].

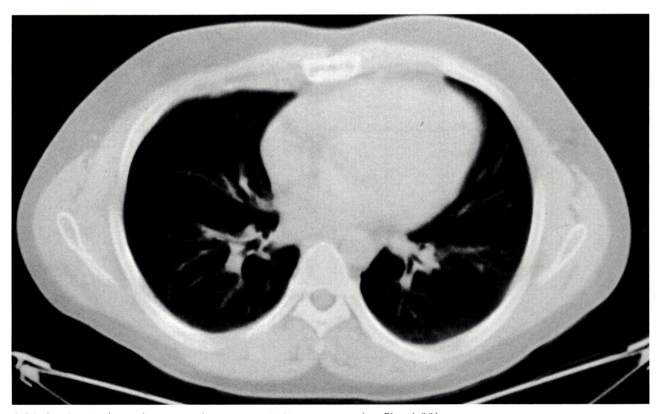

4-31 Section 9 [from slice 7 on drawing (p. 118) accompanying Fig. 4-22].

4/Chest, Computed Tomography

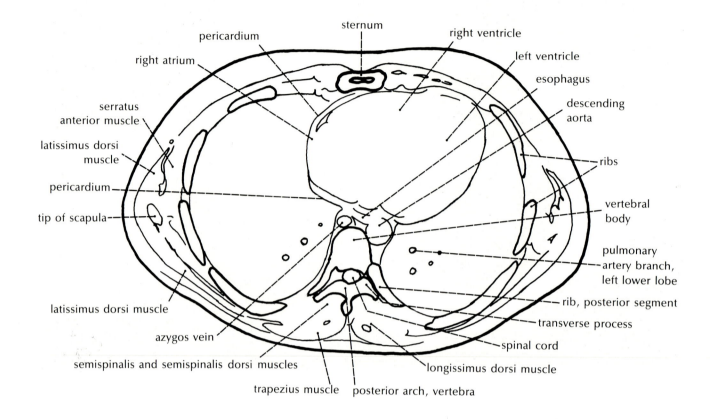

sternum

pericardium

right ventricle

right atrium

left ventricle

esophagus

serratus
anterior muscle

descending
aorta

latissimus dorsi
muscle

ribs

pericardium

tip of scapula

vertebral
body

pulmonary
artery branch,
left lower lobe

latissimus dorsi muscle

rib, posterior segment

azygos vein

transverse process

spinal cord

semispinalis and semispinalis dorsi muscles

longissimus dorsi muscle

trapezius muscle   posterior arch, vertebra

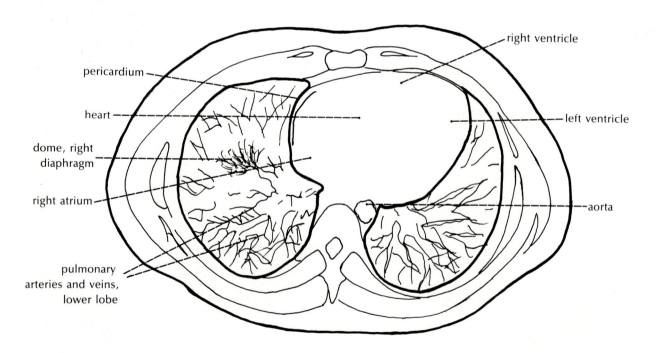

right ventricle

pericardium

heart

left ventricle

dome, right
diaphragm

right atrium

aorta

pulmonary
arteries and veins,
lower lobe

4/Chest, Computed Tomography

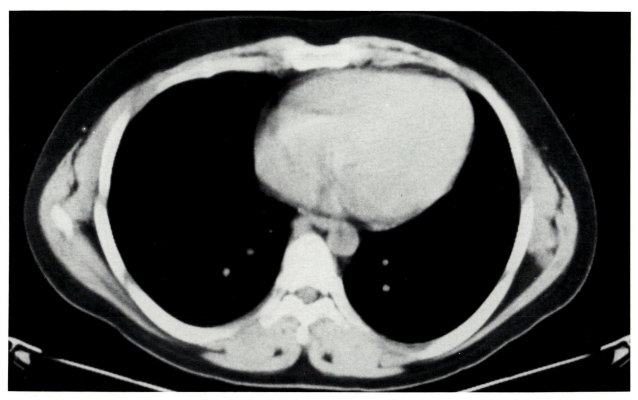

4-32 Section 10 [from slice 8 on drawing (p. 118) accompanying Fig. 4-22].

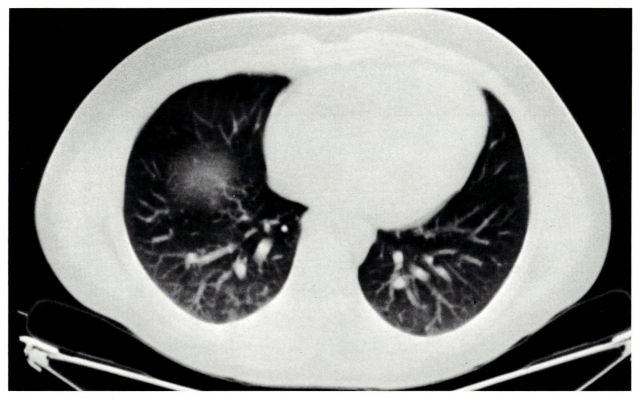

4-33 Section 11 [from slice 8 on drawing (p. 118) accompanying Fig. 4-22].

4/Chest, Computed Tomography

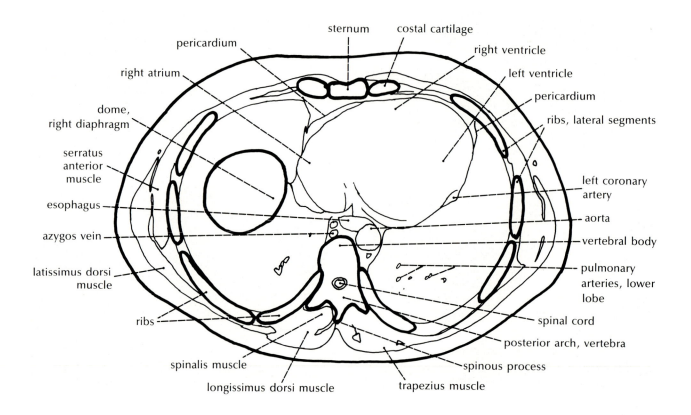

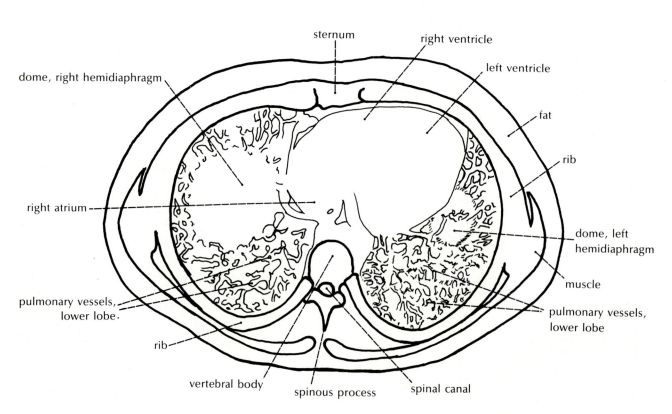

4/Chest, Computed Tomography

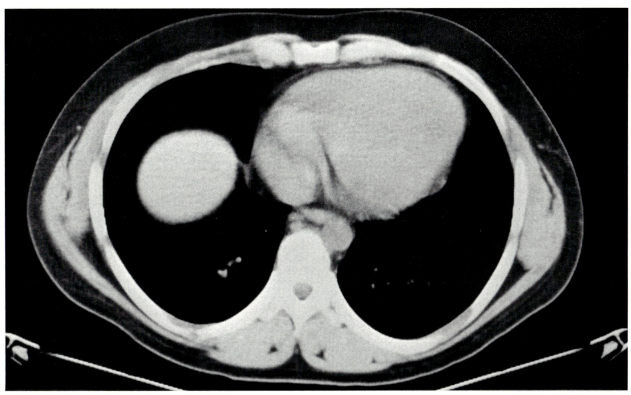

4-34 Section 12 [from slice 9 on drawing (p. 118) accompanying Fig. 4-22].

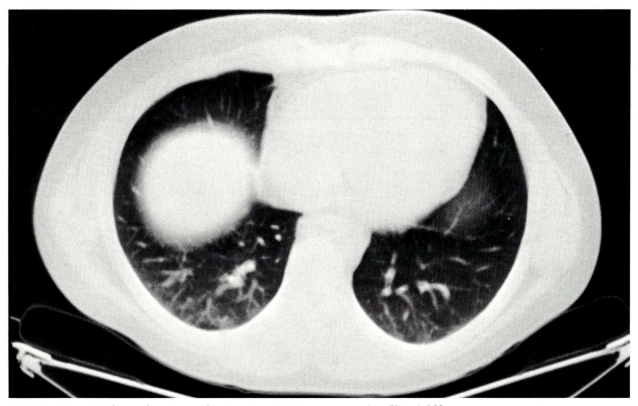

4-35 Section 13 [from slice 9 on drawing (p. 118) accompanying Fig. 4-22].

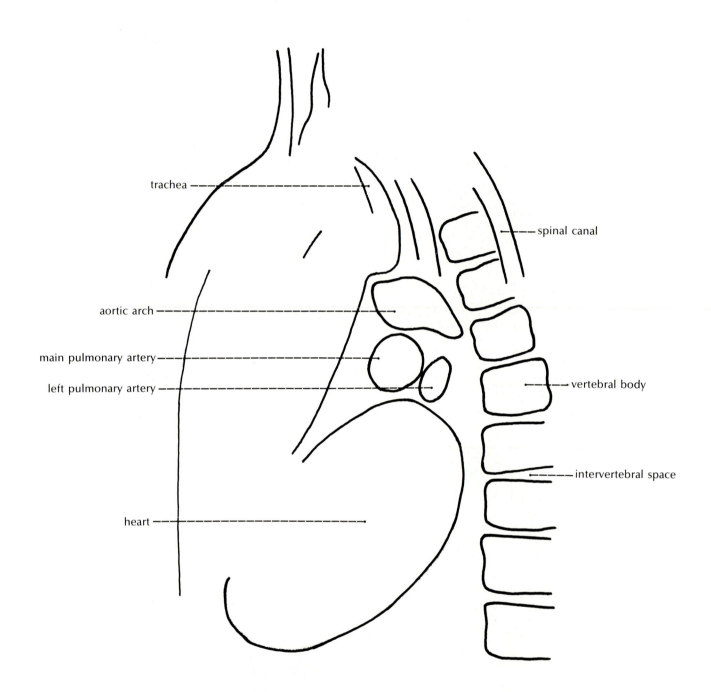

trachea

spinal canal

aortic arch

main pulmonary artery

left pulmonary artery

vertebral body

intervertebral space

heart

4-36 Magnetic resonance image (MRI), sagittal section of the chest.

4/Chest, MRI

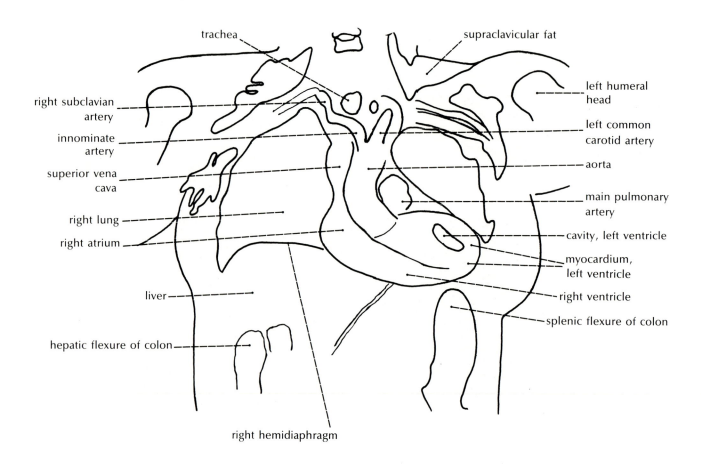

trachea

supraclavicular fat

left humeral head

right subclavian artery

innominate artery

left common carotid artery

superior vena cava

aorta

right lung

main pulmonary artery

right atrium

cavity, left ventricle

myocardium, left ventricle

liver

right ventricle

splenic flexure of colon

hepatic flexure of colon

right hemidiaphragm

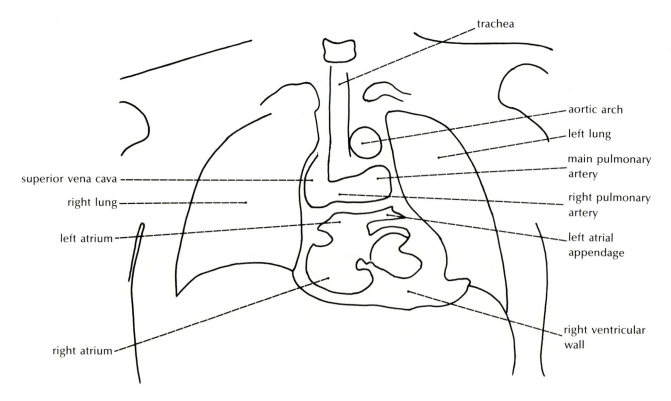

trachea

aortic arch

left lung

main pulmonary artery

superior vena cava

right lung

right pulmonary artery

left atrium

left atrial appendage

right atrium

right ventricular wall

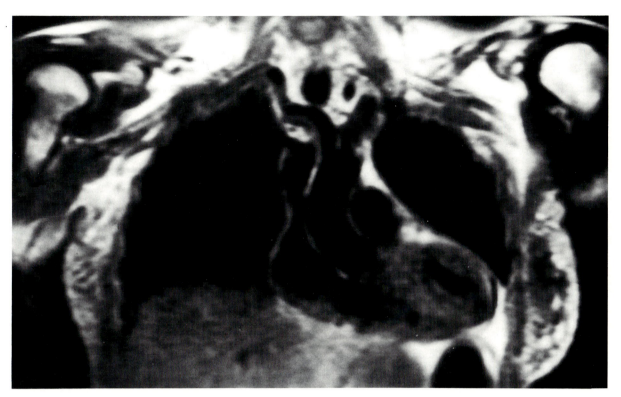

4-37 MRI, frontal section, posterior of the chest.

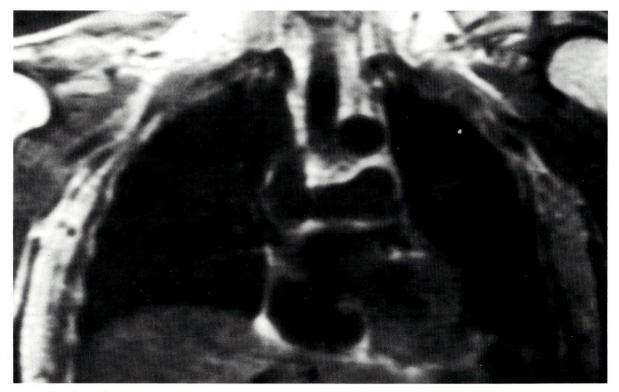

4-38 MRI, frontal section, anterior of the chest.

4/Chest, MRI

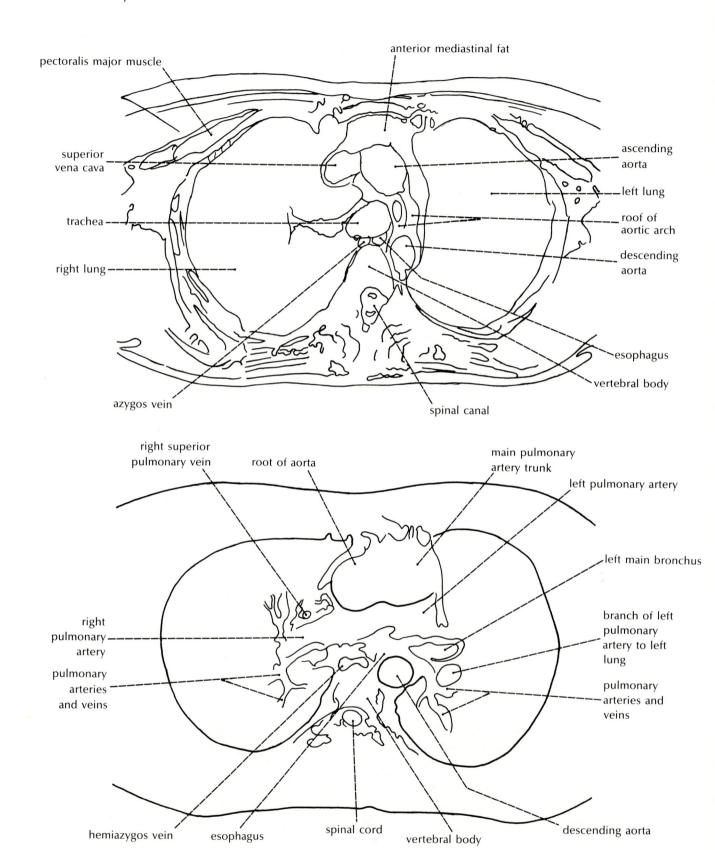

pectoralis major muscle

anterior mediastinal fat

superior
vena cava

ascending
aorta

left lung

trachea

roof of
aortic arch

descending
aorta

right lung

esophagus

vertebral body

azygos vein

spinal canal

right superior
pulmonary vein

root of aorta

main pulmonary
artery trunk

left pulmonary artery

left main bronchus

right
pulmonary
artery

branch of left
pulmonary
artery to left
lung

pulmonary
arteries
and veins

pulmonary
arteries and
veins

hemiazygos vein

esophagus

spinal cord

vertebral body

descending aorta

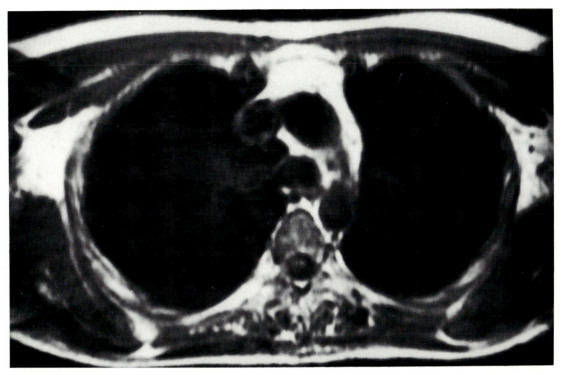

4-39 MRI, axial cross section of upper chest.

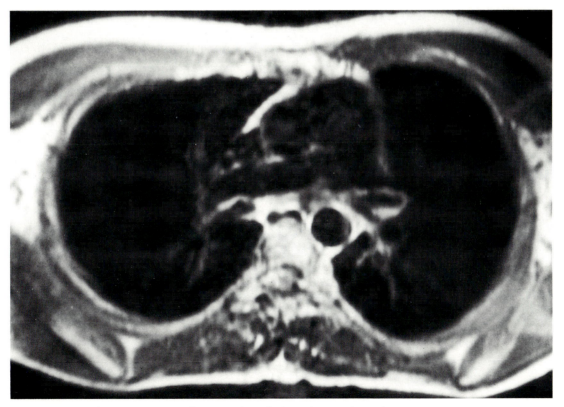

4-40 MRI, axial cross section of upper chest below the aortic arch.

4/Chest, MRI

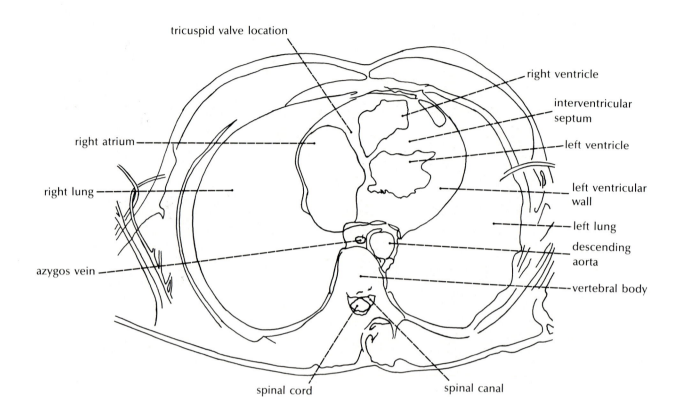

tricuspid valve location

right ventricle

interventricular septum

right atrium

left ventricle

right lung

left ventricular wall

left lung

descending aorta

azygos vein

vertebral body

spinal cord

spinal canal

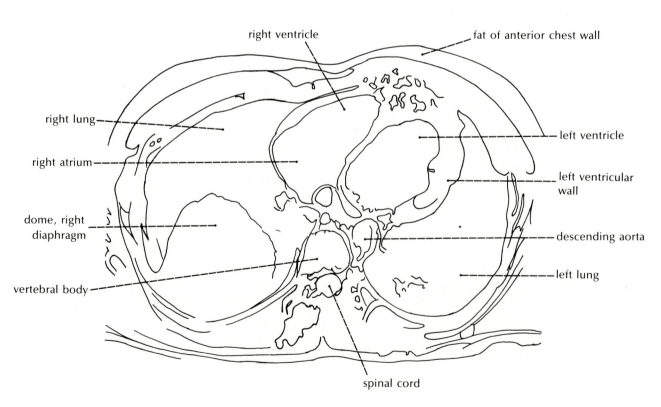

right ventricle

fat of anterior chest wall

right lung

left ventricle

right atrium

left ventricular wall

dome, right diaphragm

descending aorta

left lung

vertebral body

spinal cord

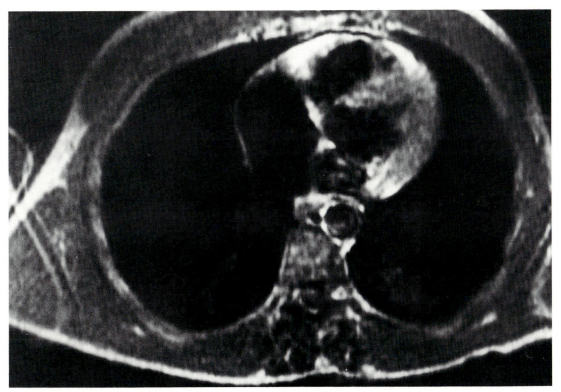

4-41 MRI, axial cross section of the chest at the level of the heart.

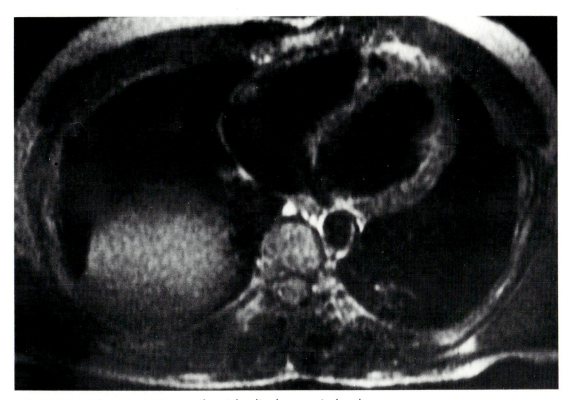

4-42 MRI, axial cross section at the right diaphragmatic level.

4/Radionuclide Lung Perfusion

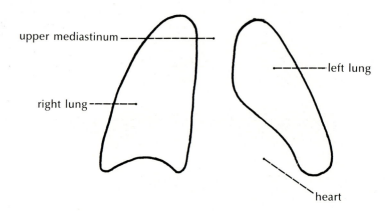

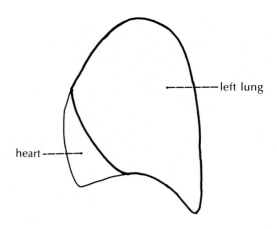

4/Radionuclide Lung Perfusion

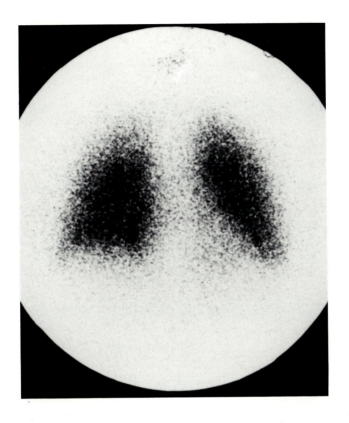

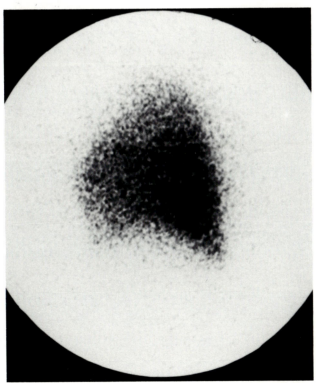

4-43 Radionuclide lung perfusion images: AP (top) and lateral (bottom).

5/Abdomen □ Stomach and Duodenum

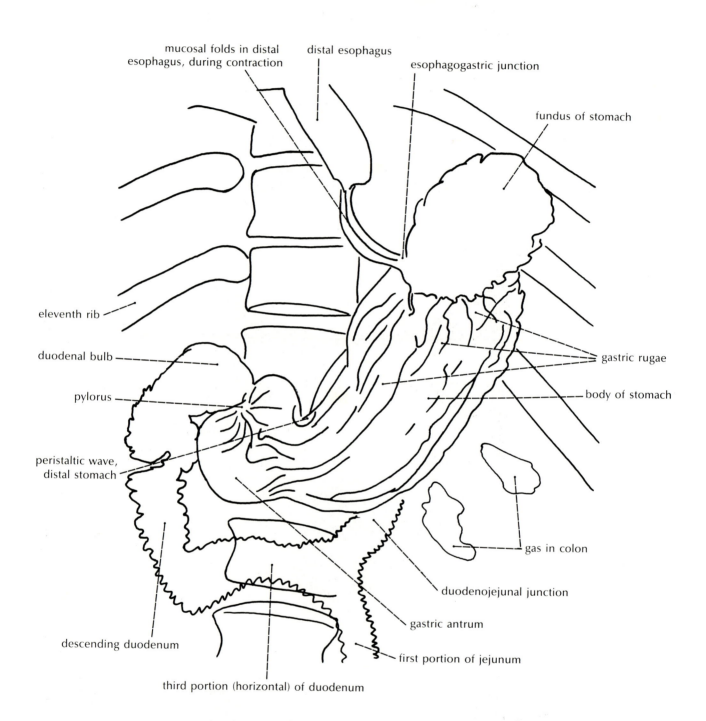

mucosal folds in distal esophagus, during contraction

distal esophagus

esophagogastric junction

fundus of stomach

eleventh rib

duodenal bulb

pylorus

gastric rugae

body of stomach

peristaltic wave, distal stomach

gas in colon

duodenojejunal junction

gastric antrum

descending duodenum

first portion of jejunum

third portion (horizontal) of duodenum

5/Stomach and Duodenum

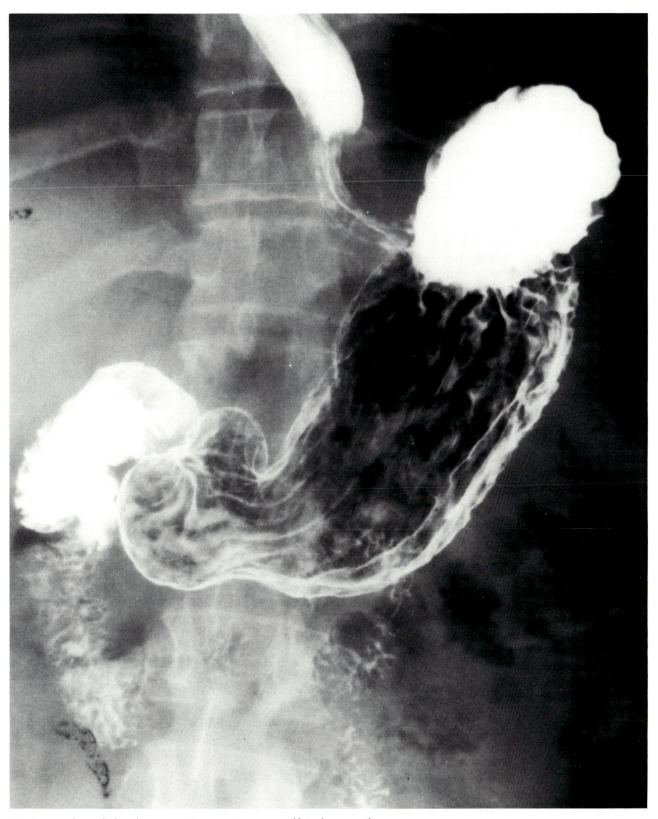

5-1 Stomach and duodenum, AP view. Barium sulfate by mouth.

5/Distal Esophagus; Stomach and Duodenum

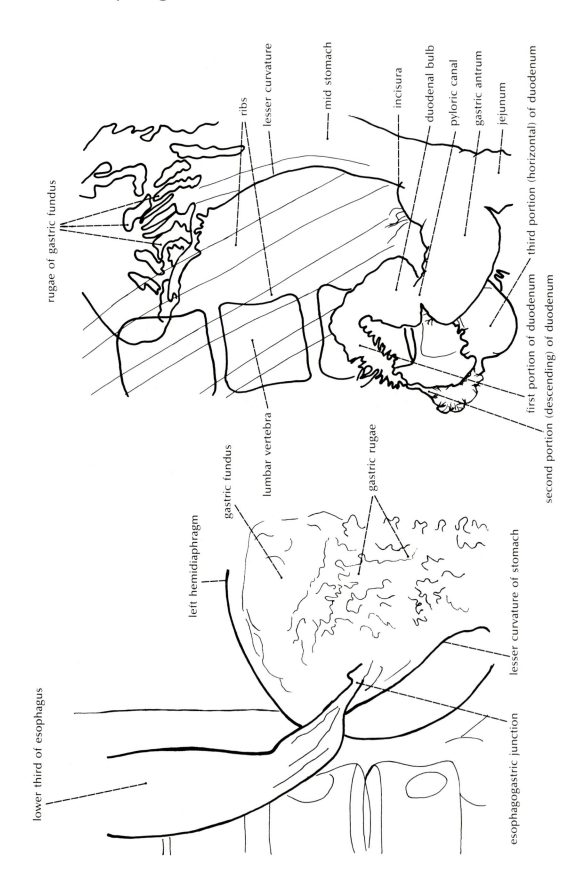

rugae of gastric fundus

ribs

lesser curvature

mid stomach

incisura

duodenal bulb

pyloric canal

gastric antrum

jejunum

first portion of duodenum

third portion (horizontal) of duodenum

second portion (descending) of duodenum

gastric fundus

lumbar vertebra

gastric rugae

left hemidiaphragm

lesser curvature of stomach

lower third of esophagus

esophagogastric junction

5/Distal Esophagus; Stomach and Duodenum

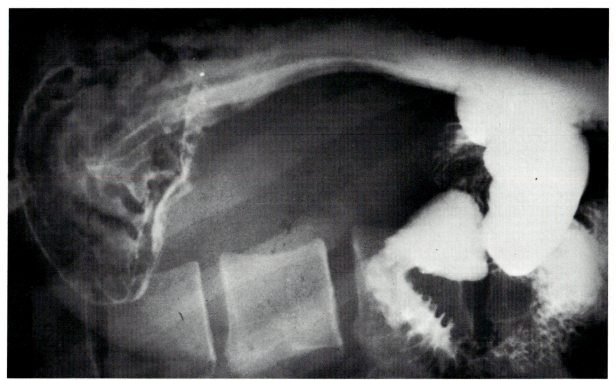

5-3 Stomach and duodenum, oblique view. Barium sulfate by mouth.

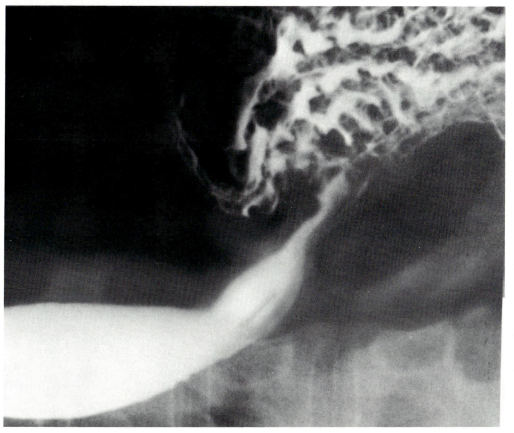

5-2 Distal esophagus, esophagogastric junction, and gastric fundus. Barium sulfate by mouth.

5/Duodenum

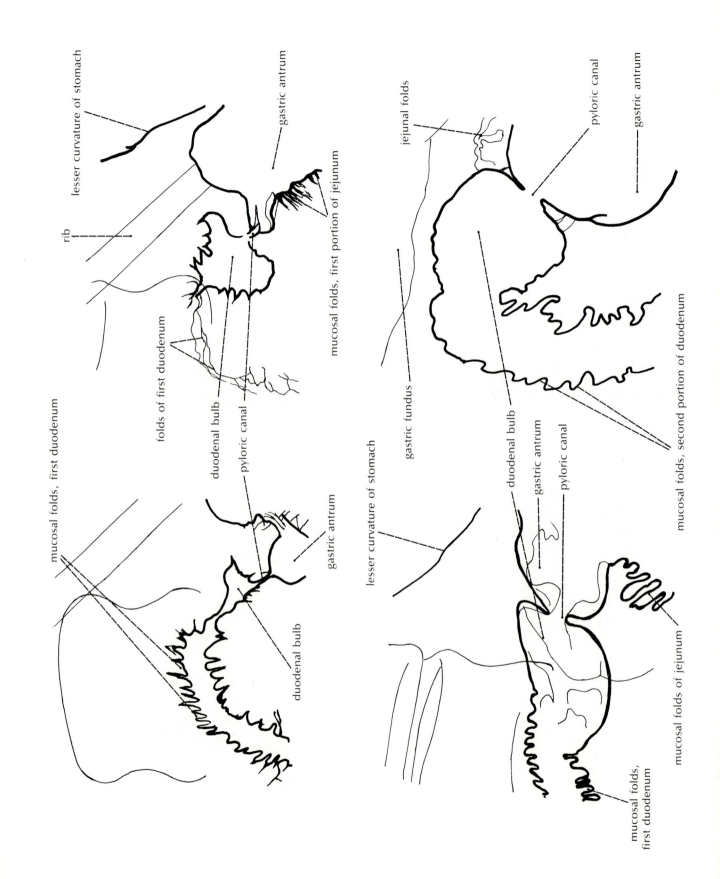

lesser curvature of stomach

gastric antrum

rib

mucosal folds, first portion of jejunum

folds of first duodenum

duodenal bulb

pyloric canal

mucosal folds, first duodenum

gastric antrum

duodenal bulb

jejunal folds

pyloric canal

gastric antrum

gastric fundus

duodenal bulb

gastric antrum

pyloric canal

mucosal folds, second portion of duodenum

lesser curvature of stomach

mucosal folds of jejunum

mucosal folds, first duodenum

5/Duodenum

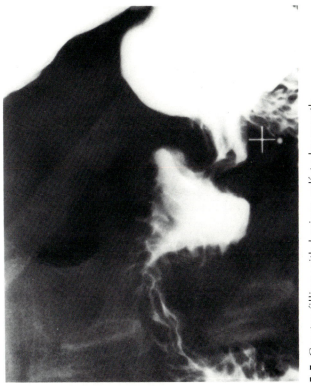

5-5 Greater filling with barium sulfate by mouth.

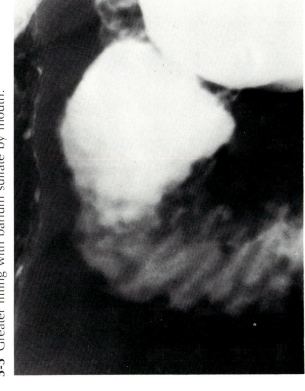

5-7 Duodenal bulb, gastric antrum, and first portion of duodenum. Maximum filling with barium sulfate by mouth.

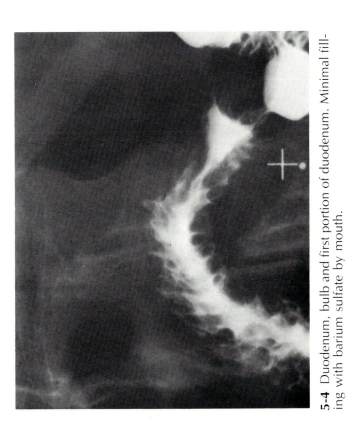

5-4 Duodenum, bulb and first portion of duodenum. Minimal filling with barium sulfate by mouth.

5-6 Duodenal bulb and gastric antrum. Air-barium contrast.

5/Jejunum and Ileum

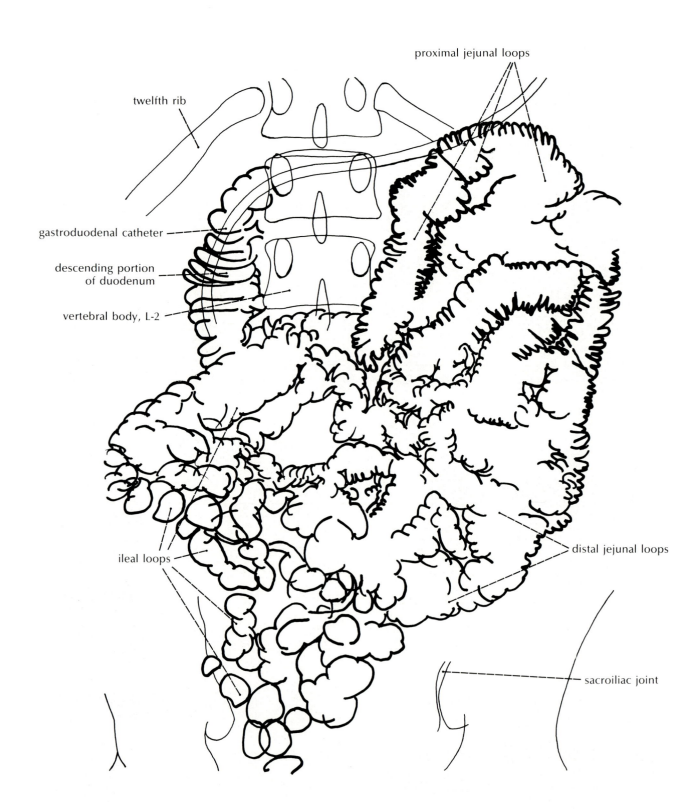

proximal jejunal loops

twelfth rib

gastroduodenal catheter

descending portion
of duodenum

vertebral body, L-2

distal jejunal loops

ileal loops

sacroiliac joint

5/Jejunum and Ileum

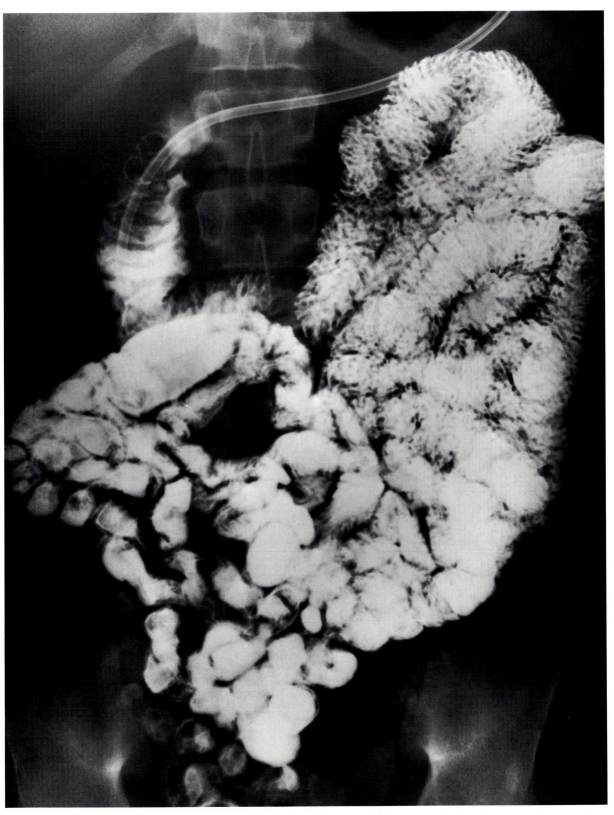

5-8 Duodenum, jejunum, and ileum. Barium sulfate administered by gastroduodenal catheter.

5/Gallbladder; Bile and Pancreatic Ducts

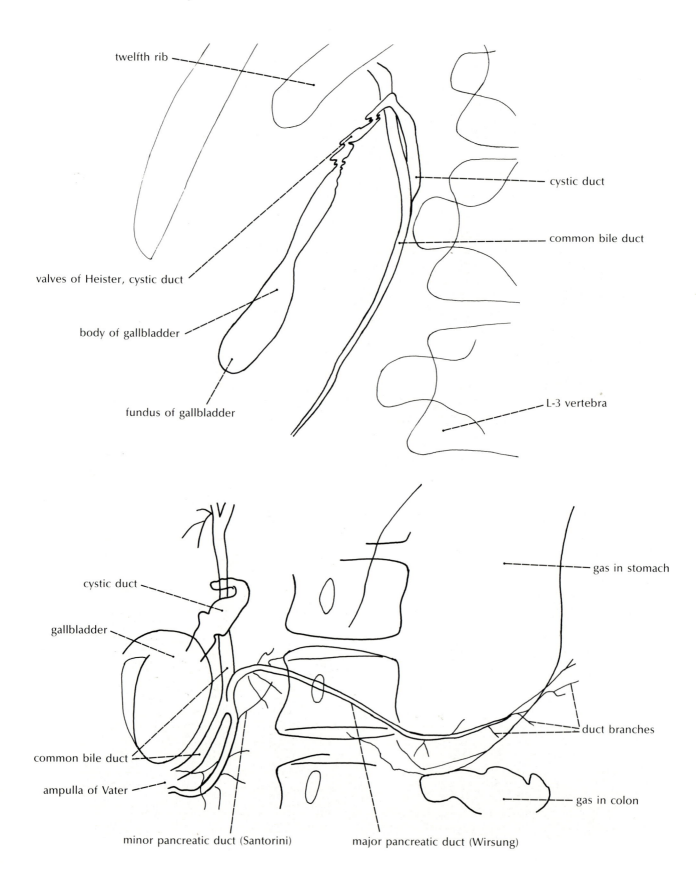

twelfth rib

cystic duct

common bile duct

valves of Heister, cystic duct

body of gallbladder

fundus of gallbladder

L-3 vertebra

cystic duct

gallbladder

gas in stomach

duct branches

common bile duct

ampulla of Vater

gas in colon

minor pancreatic duct (Santorini)

major pancreatic duct (Wirsung)

5/Gallbladder; Bile and Pancreatic Ducts

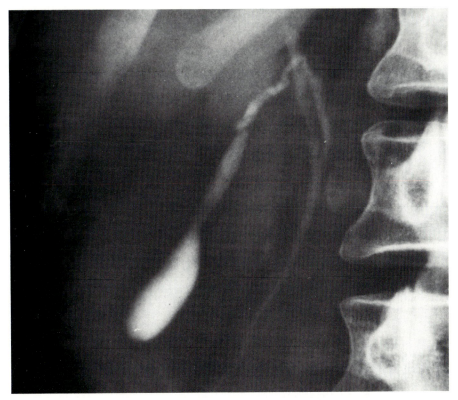

5-9 Gallbladder, cystic duct, and distal common bile duct. Visualization by oral cholecystography with gallbladder contraction.

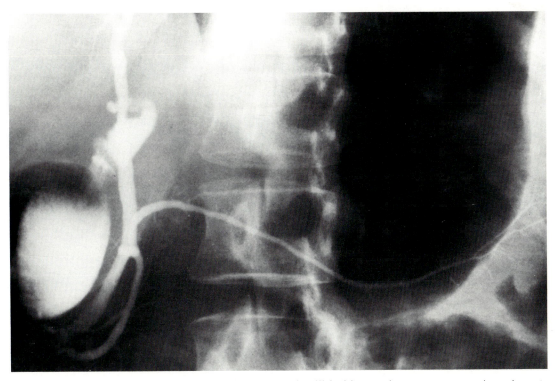

5-10 Pancreatic ducts, distal common bile duct, and gallbladder. Endoscopic retrograde catheterization of pancreatic ducts. Water-soluble contrast media injected.

5/Bile Ducts

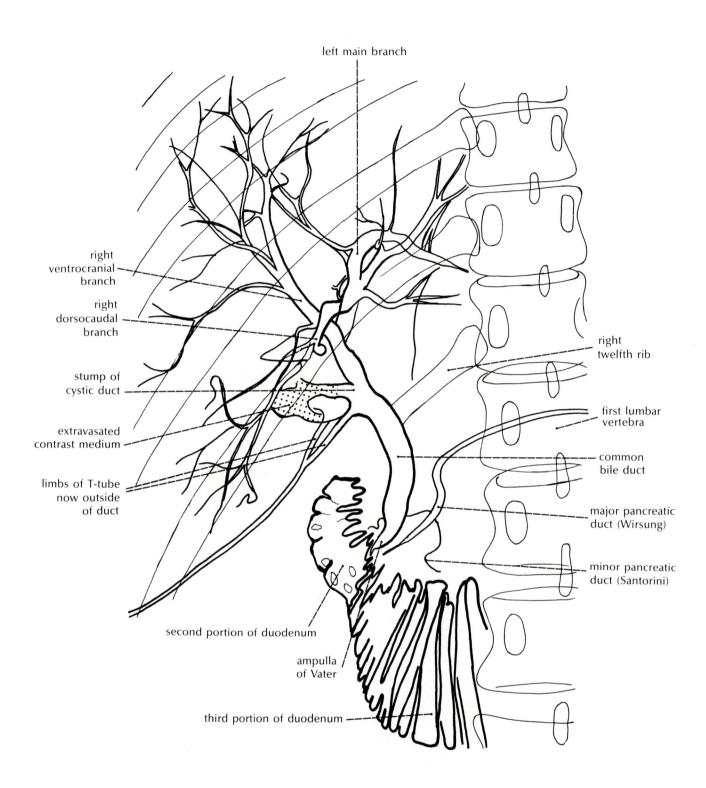

left main branch

right
ventrocranial
branch

right
dorsocaudal
branch

stump of
cystic duct

extravasated
contrast medium

limbs of T-tube
now outside
of duct

second portion of duodenum

ampulla
of Vater

third portion of duodenum

right
twelfth rib

first lumbar
vertebra

common
bile duct

major pancreatic
duct (Wirsung)

minor pancreatic
duct (Santorini)

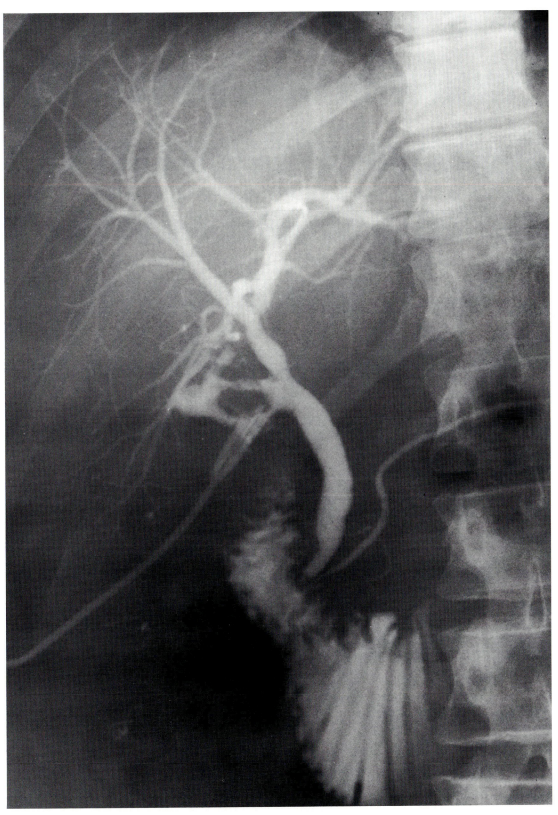

5-11 Biliary duct system, pancreatic ducts, and duodenum. Water-soluble contrast medium injected into common bile duct by an indwelling T tube inserted at time of operation. (Limbs of T-tube have come out of the common duct.)

5/Celiac Trunk, Arteriography

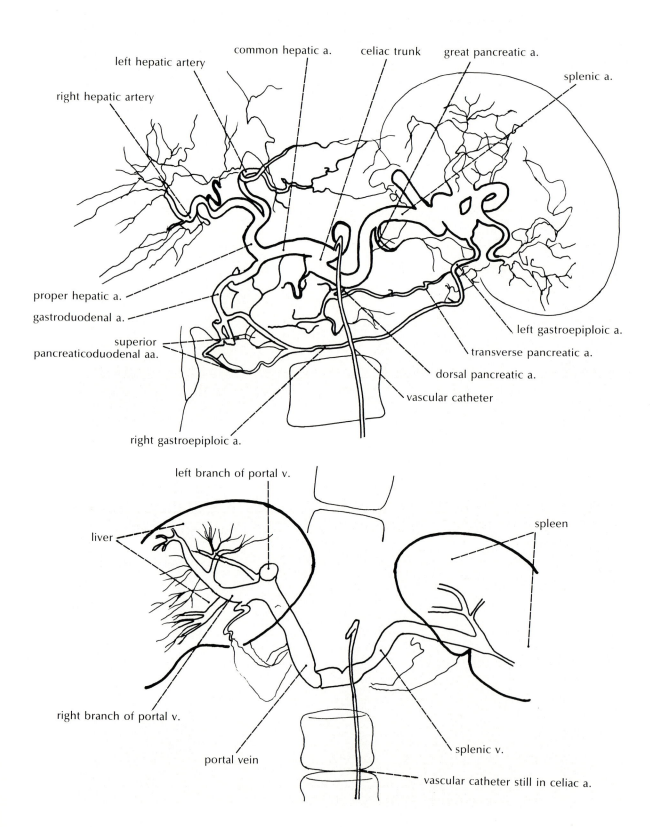

left hepatic artery

common hepatic a.

celiac trunk

great pancreatic a.

splenic a.

right hepatic artery

proper hepatic a.

gastroduodenal a.

superior
pancreaticoduodenal aa.

left gastroepiploic a.

transverse pancreatic a.

dorsal pancreatic a.

vascular catheter

right gastroepiploic a.

left branch of portal v.

liver

spleen

right branch of portal v.

portal vein

splenic v.

vascular catheter still in celiac a.

5/Celiac Trunk, Arteriography

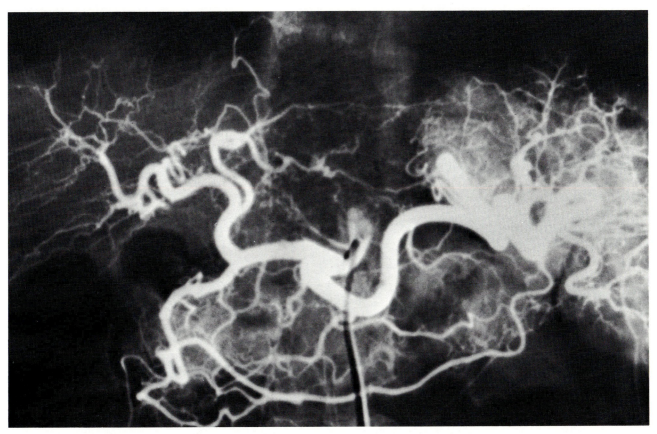

5-12 Celiac arteriogram. Water-soluble contrast medium injected into celiac trunk through a vascular catheter with tip placed in the trunk.

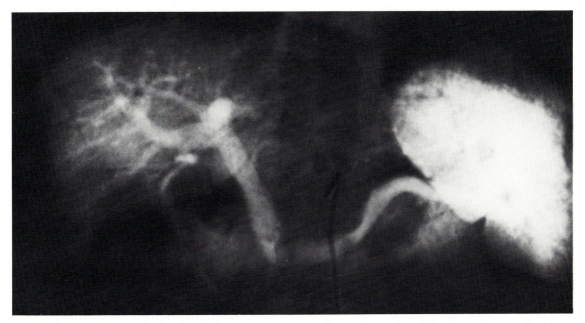

5-13 Late, or venous, phase of celiac arteriogram. Vascular catheter tip still in the celiac trunk, where water-soluble contrast had been injected.

5/Colon

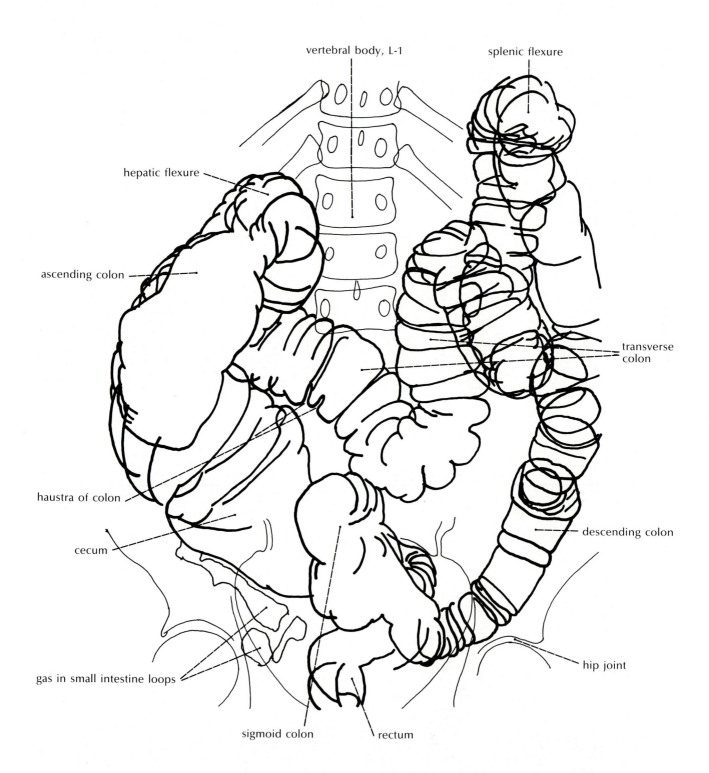

vertebral body, L-1

splenic flexure

hepatic flexure

ascending colon

transverse colon

haustra of colon

cecum

descending colon

gas in small intestine loops

hip joint

sigmoid colon

rectum

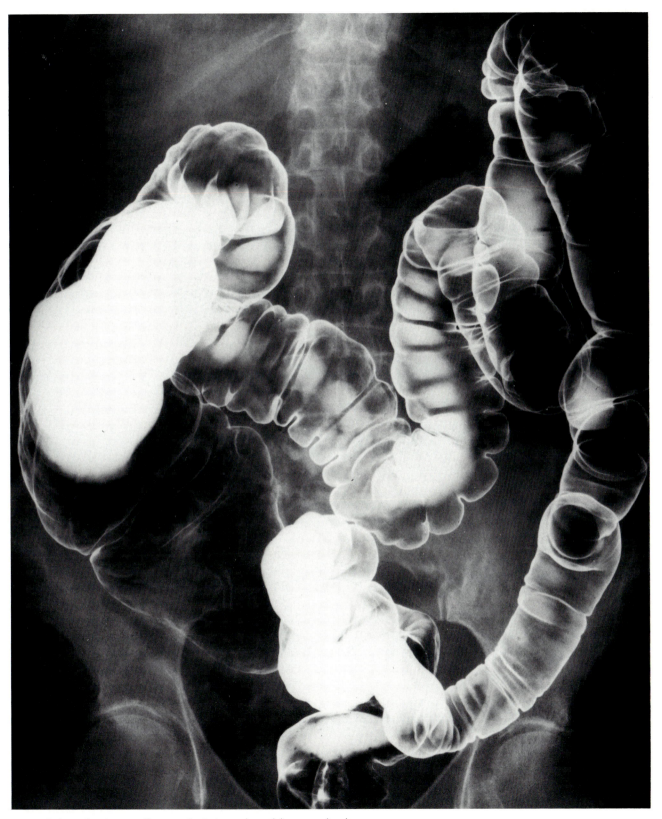

5-14 Colon, barium sulfate and air introduced by rectal tube.

5/Colon

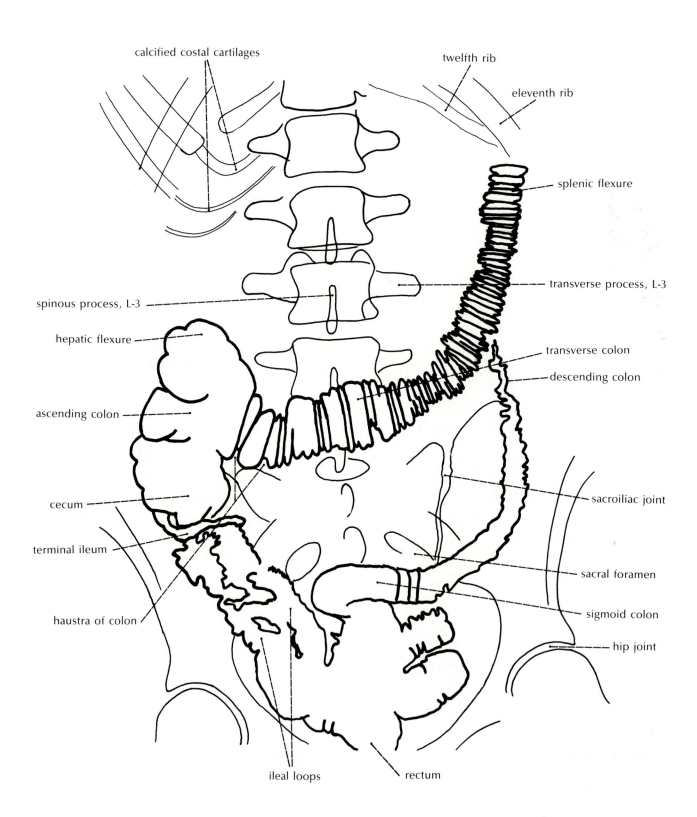

calcified costal cartilages

twelfth rib

eleventh rib

splenic flexure

transverse process, L-3

spinous process, L-3

hepatic flexure

transverse colon

descending colon

ascending colon

cecum

sacroiliac joint

terminal ileum

sacral foramen

haustra of colon

sigmoid colon

hip joint

ileal loops

rectum

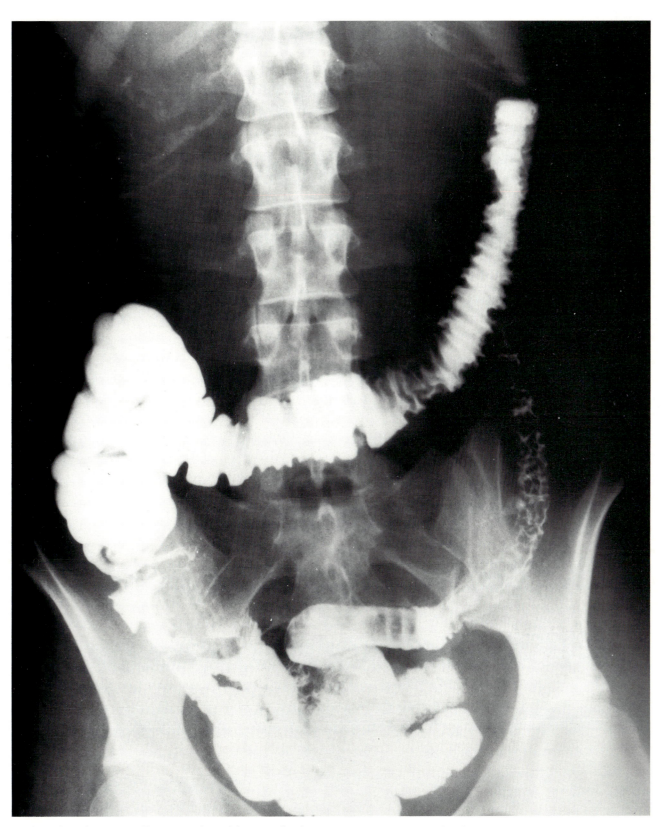

5-15 Colon, barium sulfate introduced by rectal tube. Postevacuation examination.

5/Superior Mesenteric Artery, Arteriography

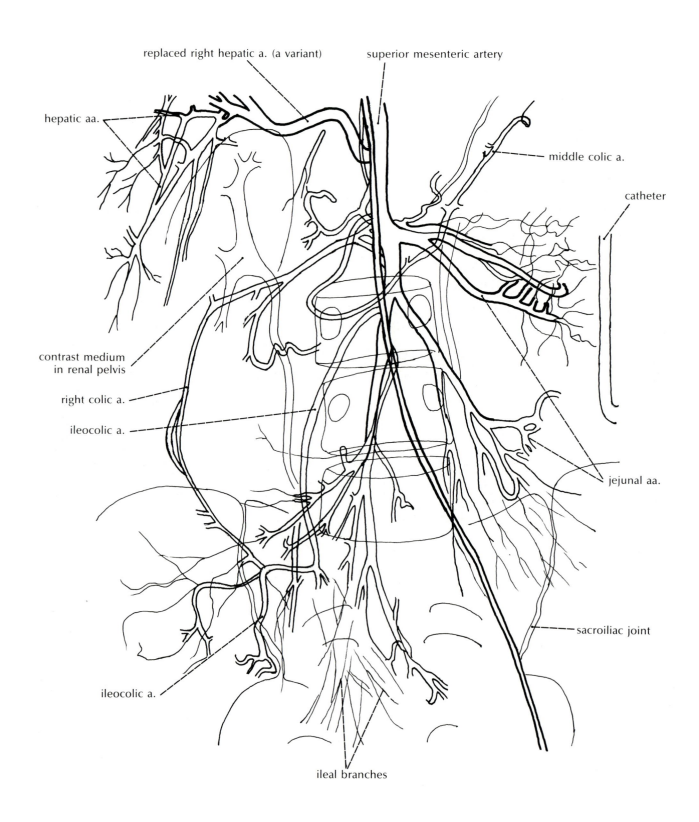

replaced right hepatic a. (a variant)

superior mesenteric artery

hepatic aa.

middle colic a.

catheter

contrast medium
in renal pelvis

right colic a.

ileocolic a.

jejunal aa.

sacroiliac joint

ileocolic a.

ileal branches

5/Superior Mesenteric Artery, Arteriography

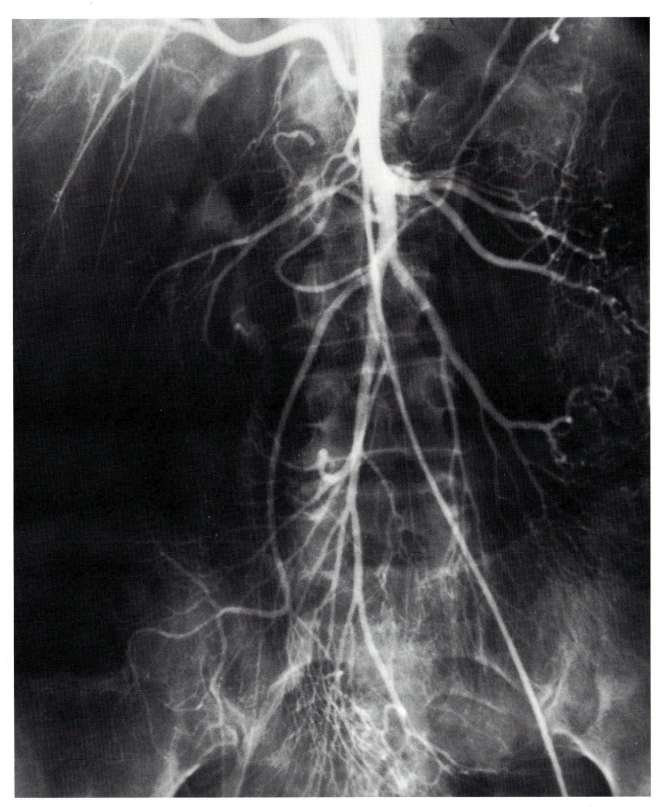

5-16 Superior mesenteric arteriogram. Water-soluble contrast medium injected through a vascular catheter introduced into root of the artery.

5/Inferior Mesenteric Artery, Arteriography

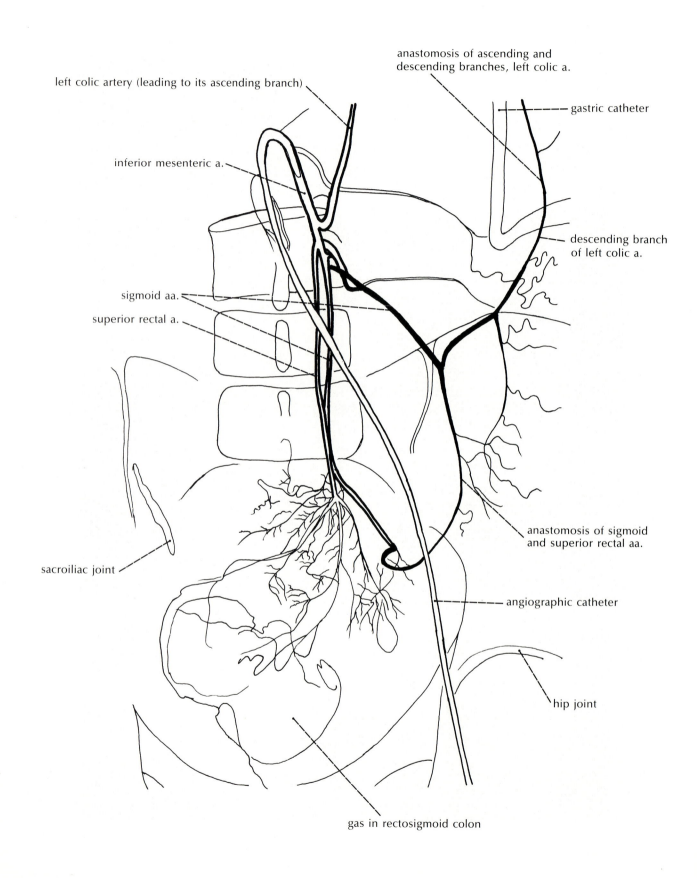

anastomosis of ascending and
descending branches, left colic a.

left colic artery (leading to its ascending branch)

gastric catheter

inferior mesenteric a.

descending branch
of left colic a.

sigmoid aa.

superior rectal a.

anastomosis of sigmoid
and superior rectal aa.

sacroiliac joint

angiographic catheter

hip joint

gas in rectosigmoid colon

5/Inferior Mesenteric Artery, Arteriography

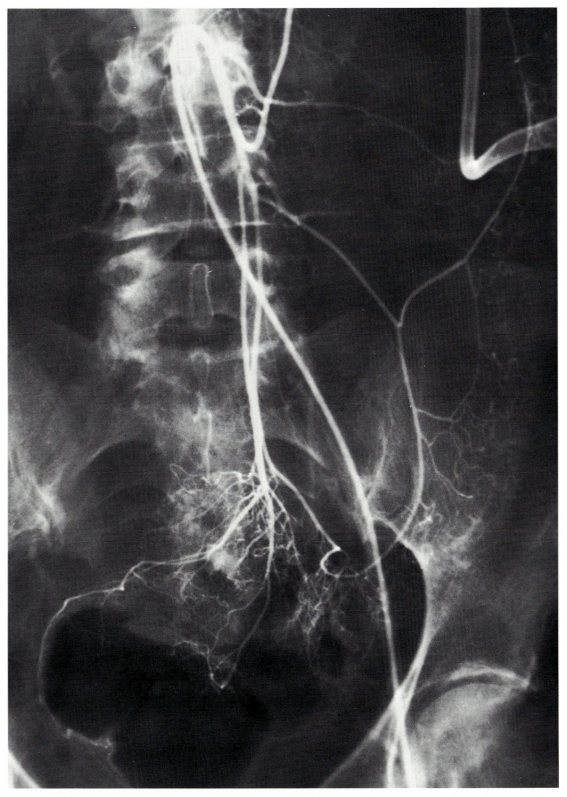

5-17 Inferior mesenteric arteriogram. Water-soluble contrast medium injected through a vascular catheter introduced into root of the artery.

5/Abdominal Aortography

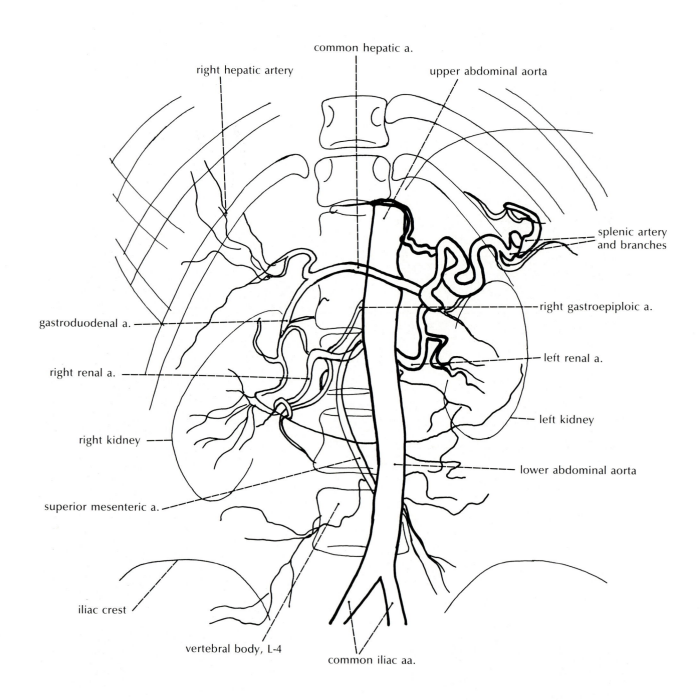

right hepatic artery

common hepatic a.

upper abdominal aorta

splenic artery and branches

gastroduodenal a.

right gastroepiploic a.

right renal a.

left renal a.

left kidney

right kidney

lower abdominal aorta

superior mesenteric a.

iliac crest

vertebral body, L-4

common iliac aa.

5/Abdominal Aortography

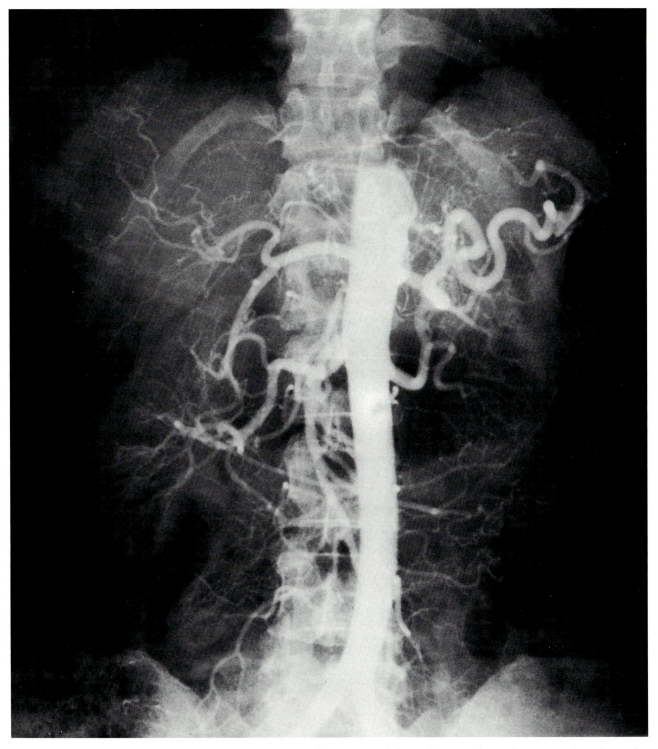

5-18 Abdominal aortogram. Water-soluble contrast medium injected through a vascular catheter introduced into the upper abdominal aorta.

5/Venacavography

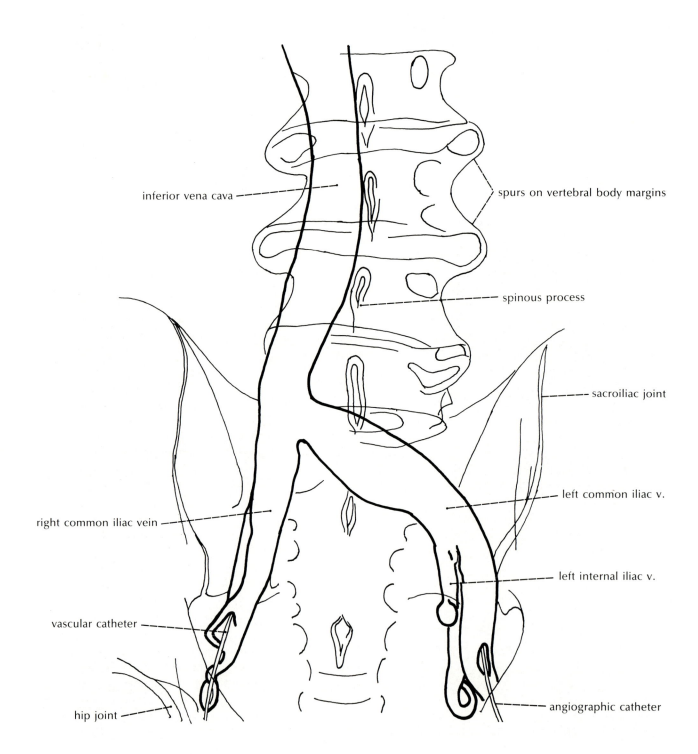

inferior vena cava

spurs on vertebral body margins

spinous process

sacroiliac joint

left common iliac v.

right common iliac vein

left internal iliac v.

vascular catheter

angiographic catheter

hip joint

5/Venacavography

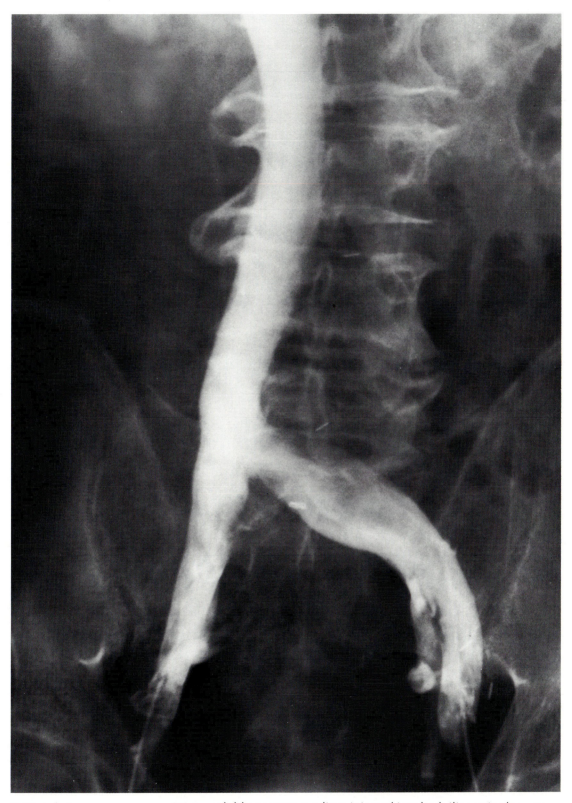

5-19 Inferior venacavagram. Water-soluble contrast medium injected into both iliac veins by vascular catheters. (Degenerative spurs on vertebral bodies in this elderly male.)

5/Urography

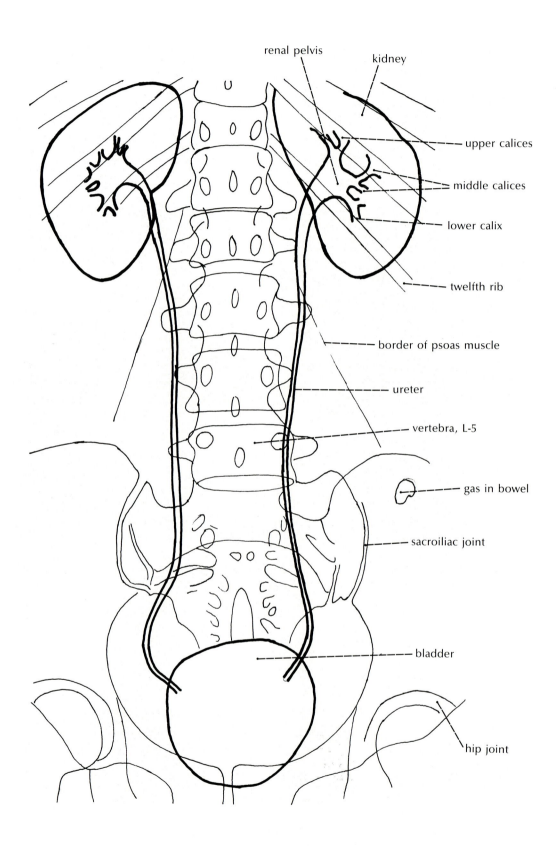

renal pelvis

kidney

upper calices

middle calices

lower calix

twelfth rib

border of psoas muscle

ureter

vertebra, L-5

gas in bowel

sacroiliac joint

bladder

hip joint

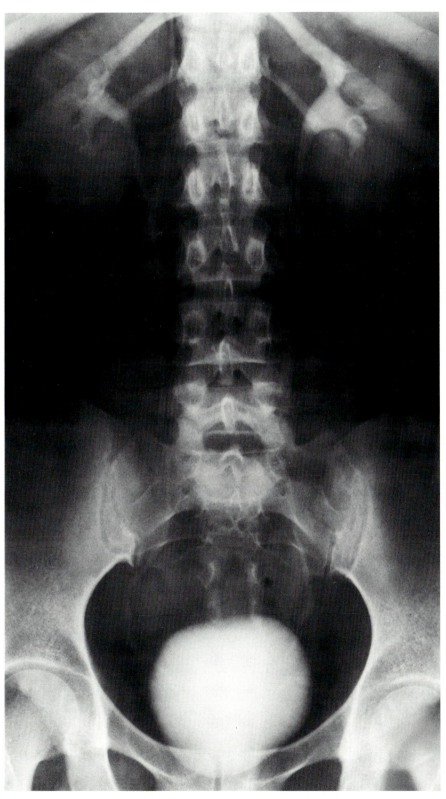

5-20 Kidneys, ureters, and bladder. Visualization by water-soluble contrast medium injected intravenously and excreted by the kidneys.

5/Posterior Abdomen, MRI

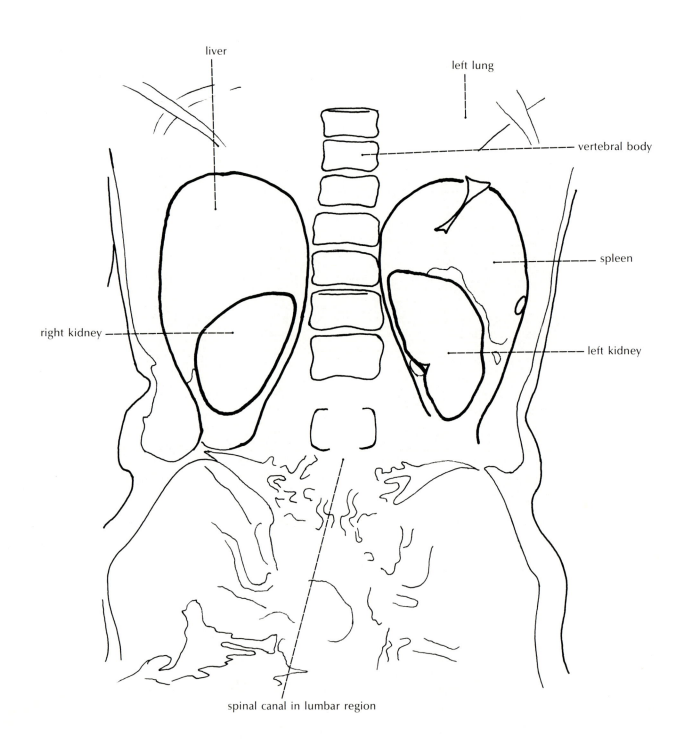

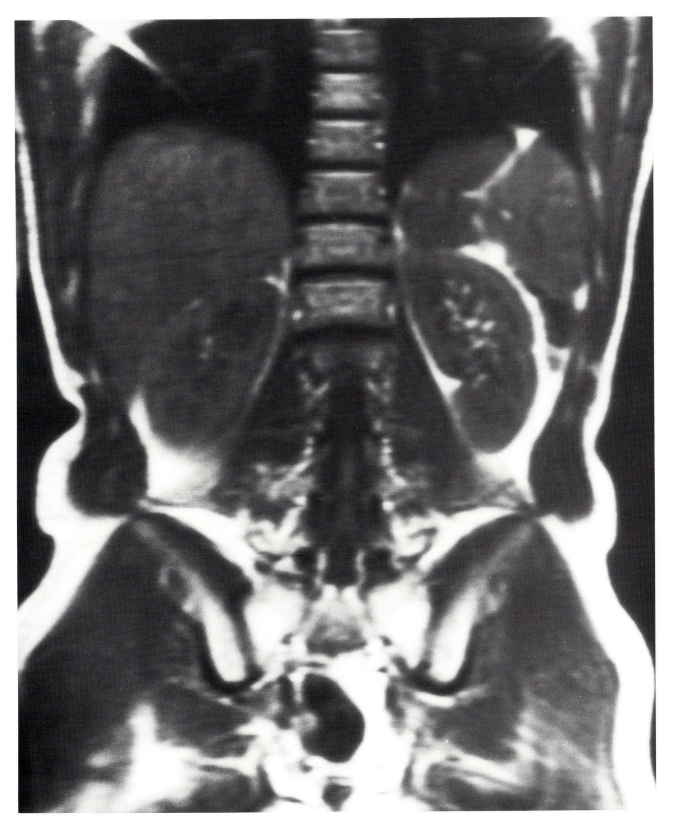

5-21 Frontal view of posterior abdomen and retroperitoneal structures. MR image.

5/Urography; Renal Arteriography

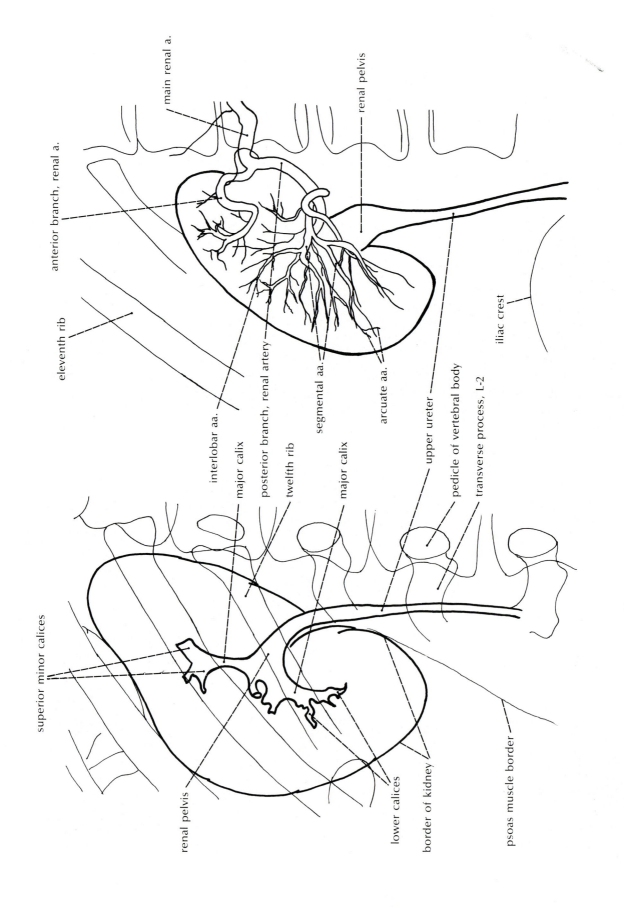

main renal a.

renal pelvis

anterior branch, renal a.

eleventh rib

interlobar aa.

major calix

posterior branch, renal artery

twelfth rib

segmental aa.

major calix

arcuate aa.

upper ureter

pedicle of vertebral body

transverse process, L-2

iliac crest

superior minor calices

renal pelvis

lower calices

border of kidney

psoas muscle border

5/Urography; Renal Arteriography

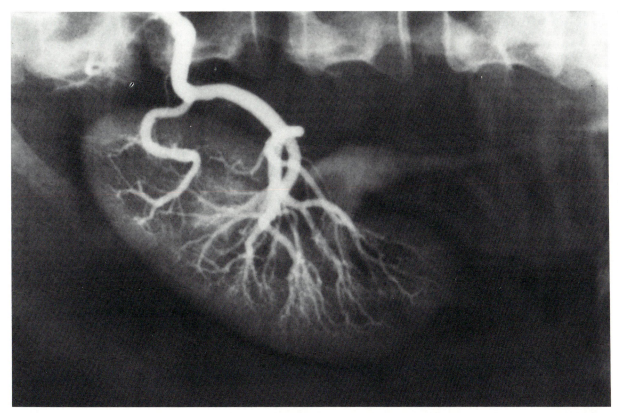

5-23 Renal arteriogram. Water-soluble contrast medium injected through a vascular catheter placed in the right renal artery.

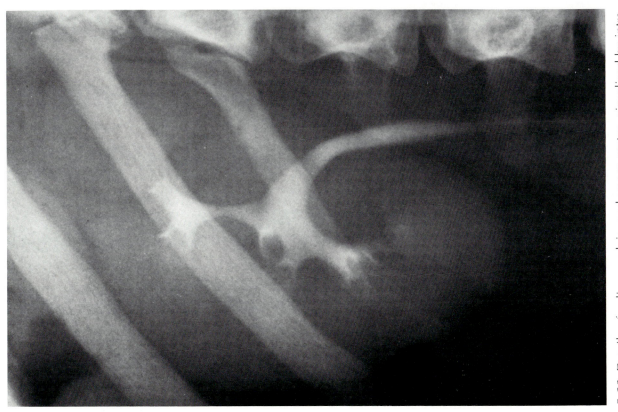

5-22 Details of calices, pelvis, and upper ureter, visualized by intravenous urography (see Fig. 5-20).

5/Adrenal Glands, CT; Urinary Tract, Isotope Scan

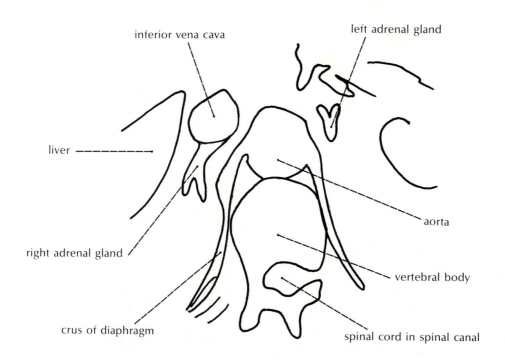

inferior vena cava

left adrenal gland

liver

right adrenal gland

crus of diaphragm

aorta

vertebral body

spinal cord in spinal canal

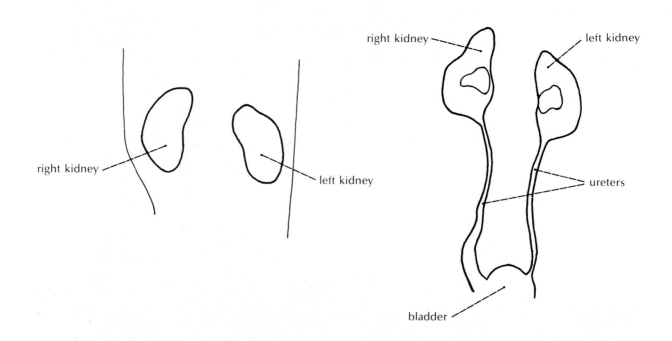

right kidney

left kidney

right kidney

left kidney

ureters

bladder

5/Adrenal Glands, CT; Urinary Tract, Isotope Scan

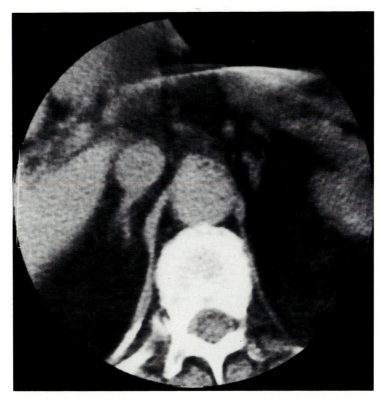

5-24 Adrenal glands. Enlargement of CT image.

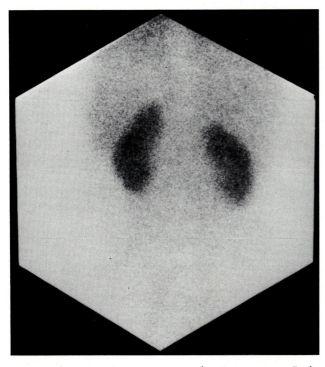

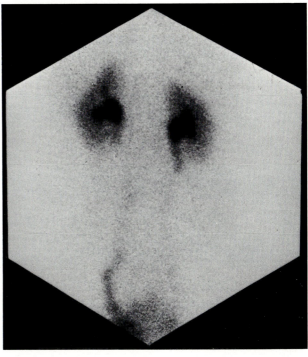

5-25 Radioactive isotope scan of urinary tract. Early phase.

5-26 Radioactive isotope scan of urinary tract. Later phase.

5/Lymphography

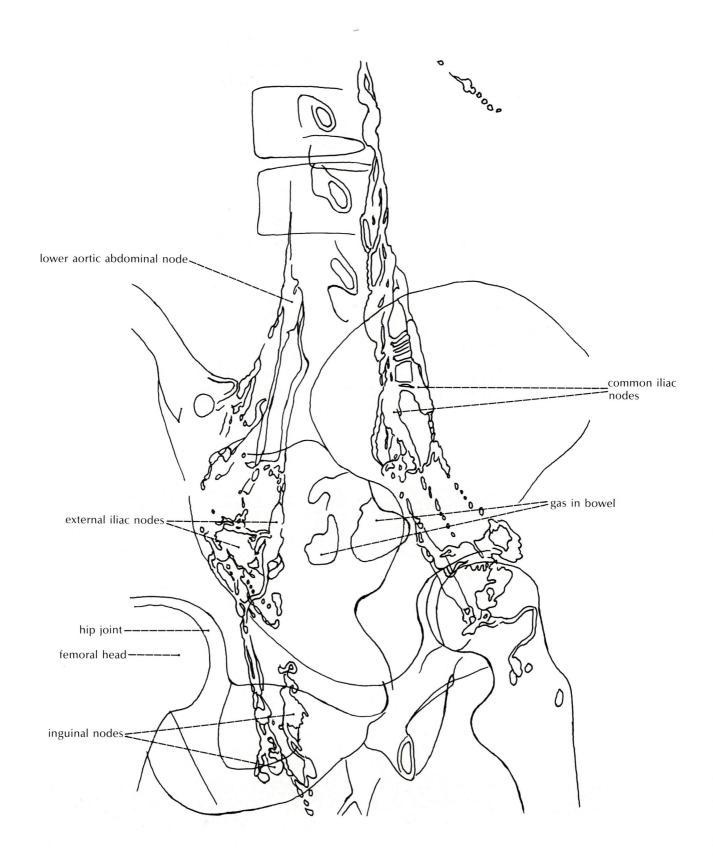

lower aortic abdominal node

common iliac
nodes

gas in bowel

external iliac nodes

hip joint

femoral head

inguinal nodes

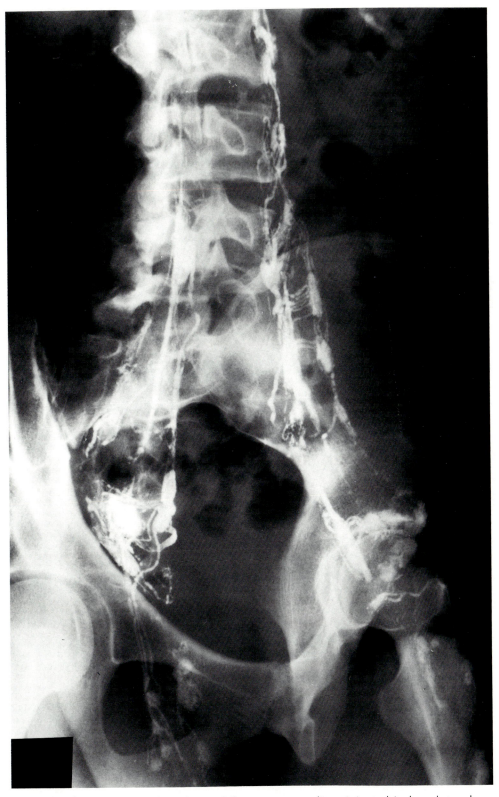

5-27 Lymphogram, injection phase. Oily contrast medium injected in lymph trunk on the dorsum of the foot.

5/Lymphography

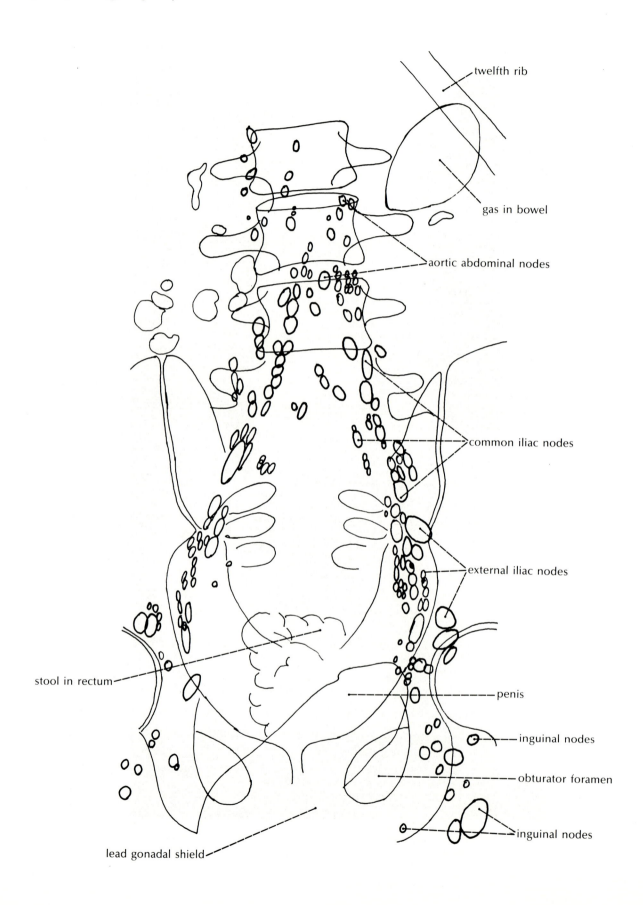

twelfth rib

gas in bowel

aortic abdominal nodes

common iliac nodes

external iliac nodes

stool in rectum

penis

inguinal nodes

obturator foramen

inguinal nodes

lead gonadal shield

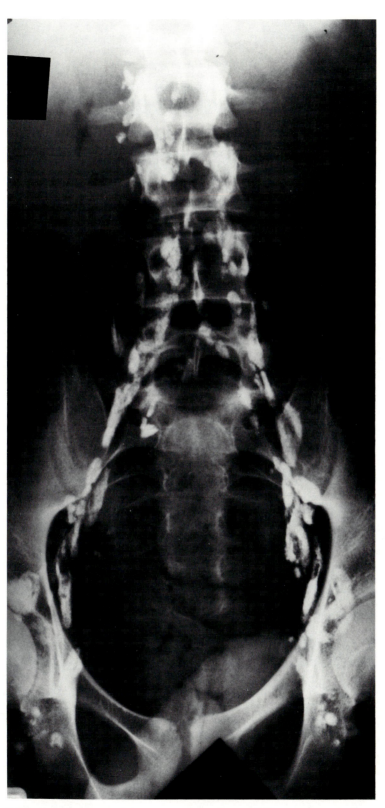

5-28 Lymphogram, later, or nodal, phase. Oily contrast medium injected in lymph trunk on the dorsum of the foot.

5/Lymphography

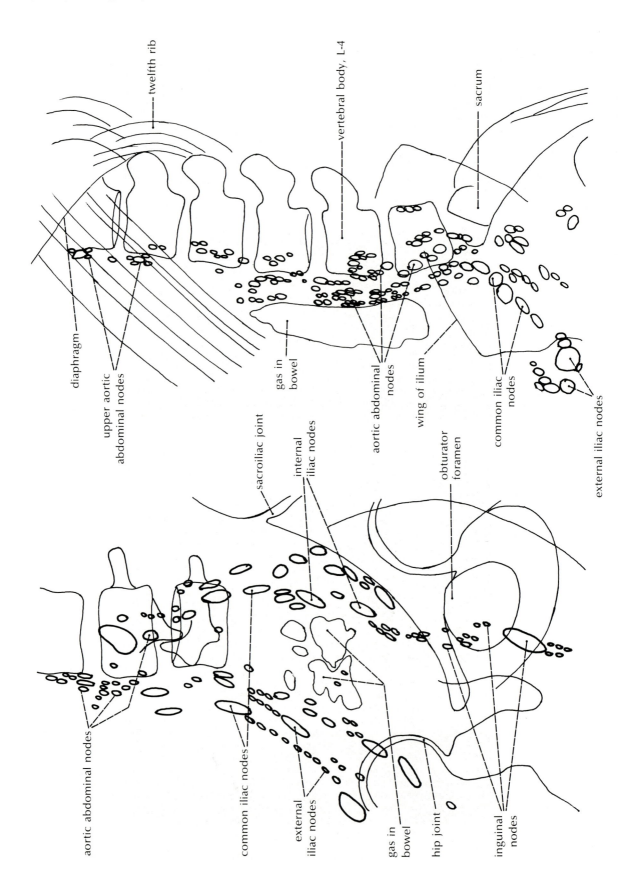

twelfth rib

vertebral body, L-4

sacrum

diaphragm

upper aortic
abdominal nodes

gas in
bowel

aortic abdominal
nodes

wing of ilium

common iliac
nodes

external iliac nodes

sacroiliac joint

internal
iliac nodes

obturator
foramen

aortic abdominal nodes

common iliac nodes

external
iliac nodes

gas in
bowel

hip joint

inguinal
nodes

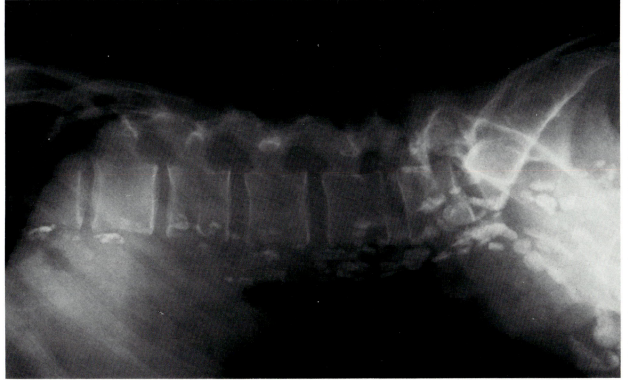

5-30 Lymphogram, later, or nodal, phase. Lateral view.

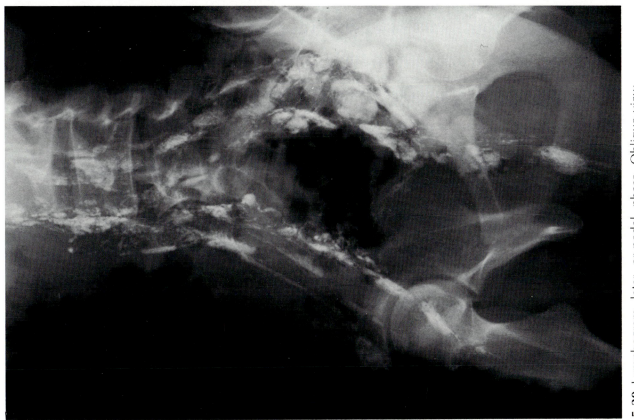

5-29 Lymphogram, later, or nodal, phase. Oblique view.

5/Cystourethrography

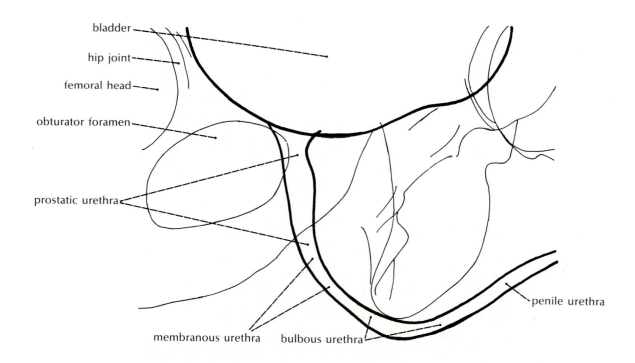

bladder

hip joint

femoral head

obturator foramen

prostatic urethra

penile urethra

membranous urethra   bulbous urethra

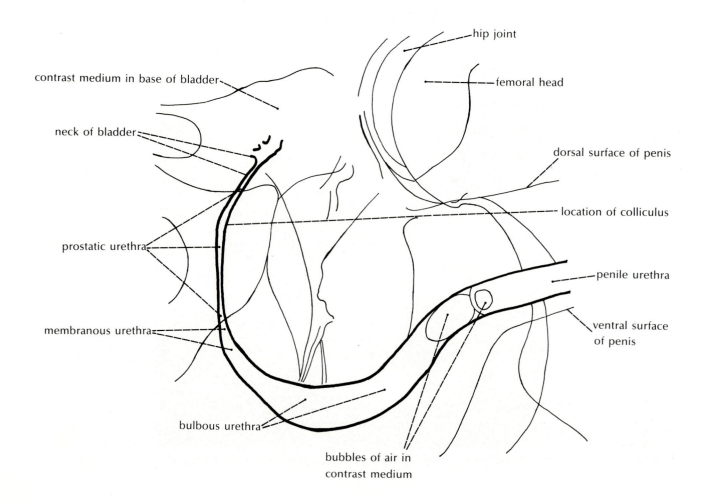

hip joint

contrast medium in base of bladder

femoral head

neck of bladder

dorsal surface of penis

location of colliculus

prostatic urethra

penile urethra

membranous urethra

ventral surface of penis

bulbous urethra

bubbles of air in contrast medium

5/Cystourethrography

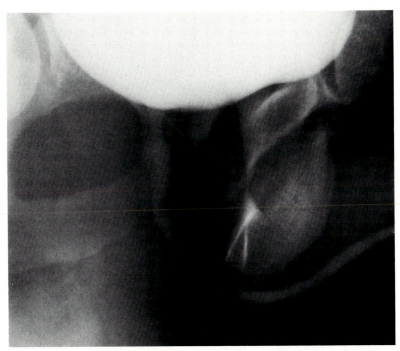

5-31 Voiding cystourethrogram in the adult male. Water-soluble contrast medium introduced into the bladder. Oblique view.

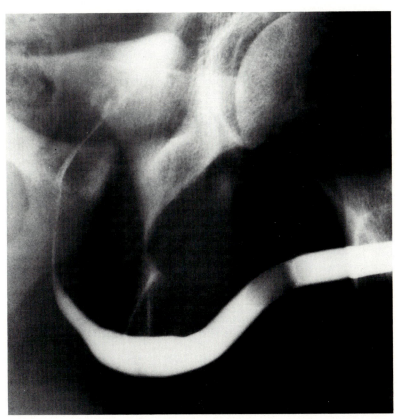

5-32 Injection cystourethrogram in the adult male. Viscous water-soluble contrast medium injected into the penis. Oblique view.

5/Hysterosalpingography

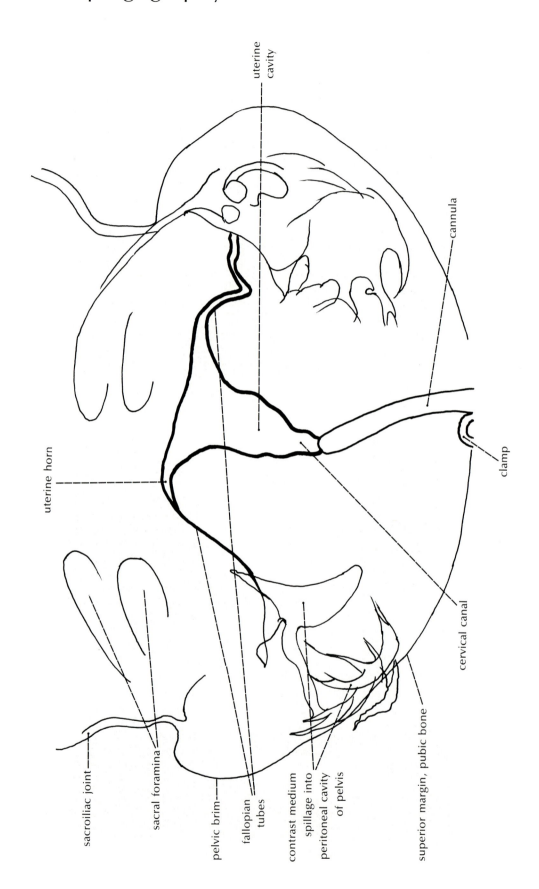

5/Hysterosalpingography

5-33 Hysterosalpingogram. Water-soluble or oily contrast medium injected via a cannula placed in the cervical canal.

5/Pelvis, Arteriography

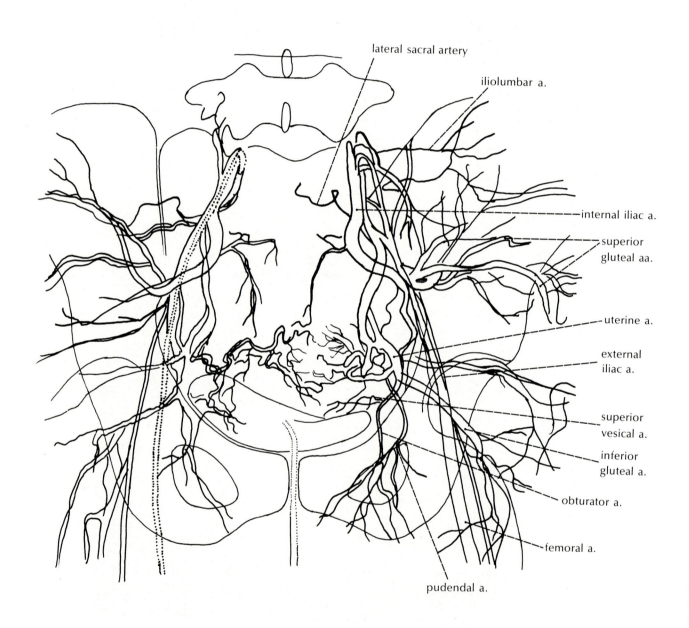

lateral sacral artery

iliolumbar a.

internal iliac a.

superior
gluteal aa.

uterine a.

external
iliac a.

superior
vesical a.

inferior
gluteal a.

obturator a.

femoral a.

pudendal a.

5/Pelvis, Arteriography

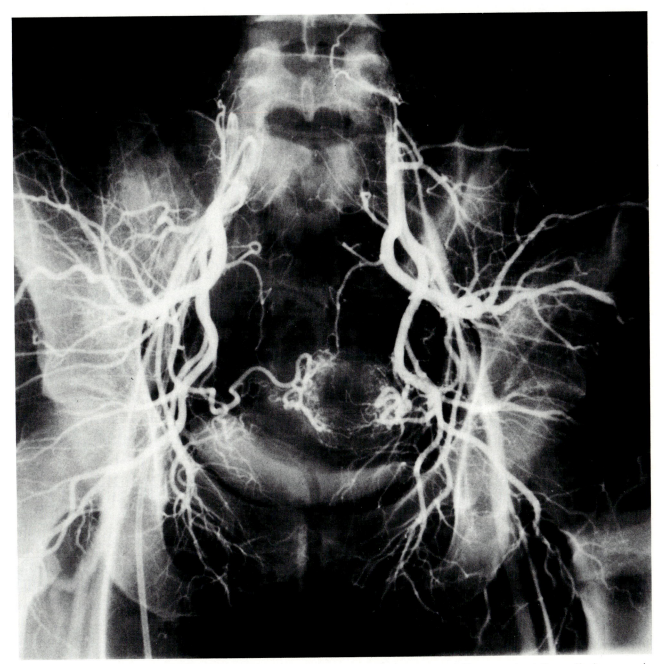

5-34 Pelvic arteriogram. Water-soluble contrast medium injected into internal iliac arteries bilaterally by vascular catheters with their tips placed in the arteries.

5/Rectosigmoid Colon and Rectum; Pelvis, MRI

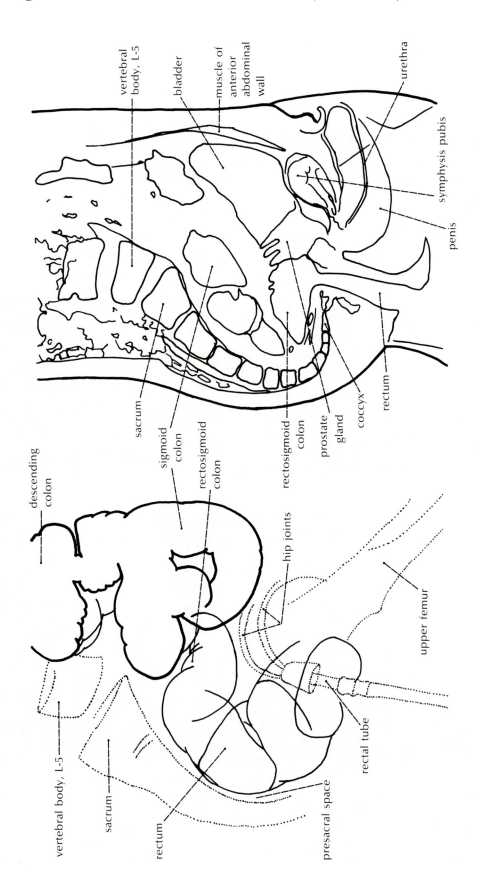

vertebral body, L-5

bladder

muscle of anterior abdominal wall

urethra

symphysis pubis

penis

sacrum

sigmoid colon

rectosigmoid colon

rectosigmoid colon

prostate gland

coccyx

rectum

descending colon

hip joints

upper femur

vertebral body, L-5

sacrum

rectum

presacral space

rectal tube

5/Rectosigmoid Colon and Rectum; Pelvis, MRI

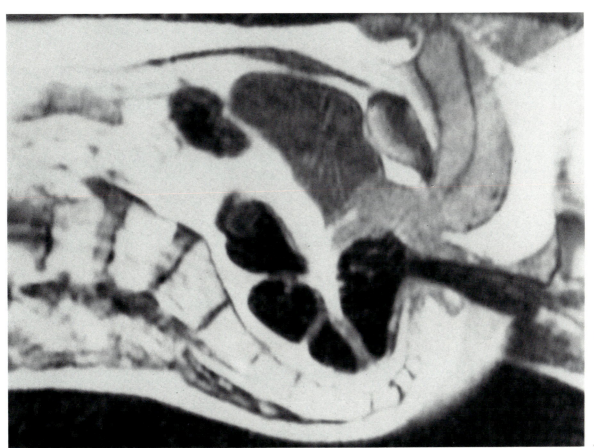

5-36 Sagittal MR image of pelvis and lower trunk.

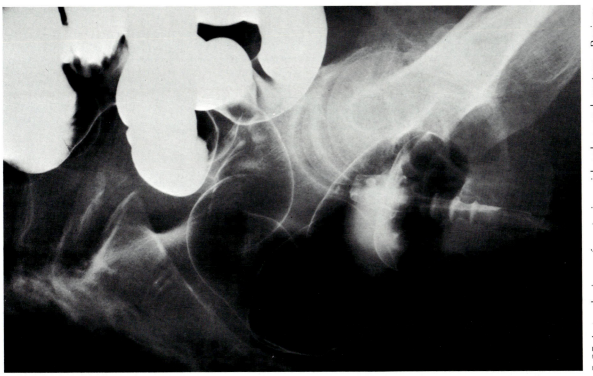

5-35 Lateral view of rectosigmoid colon and rectum. Barium sulfate introduced into rectum and colon by a cannula placed in the rectum.

5/Abdomen, Computed Tomography Sections

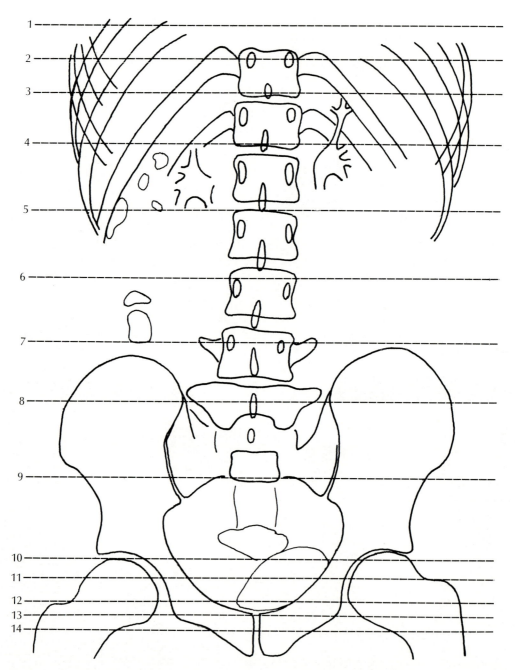

5-37 Levels of cross sections of CT scans of abdomen. Scans of four persons are utilized for these cross sections. Cross sections 11 and 13 are of the female pelvis; all other pelvic cross sections are of the male pelvis.

5/Upper Abdomen, Computed Tomography

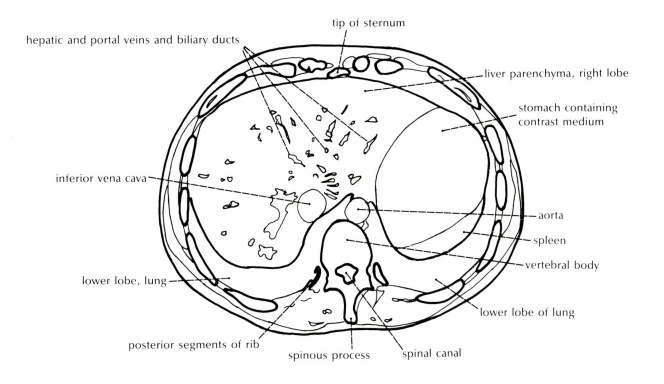

tip of sternum

hepatic and portal veins and biliary ducts

liver parenchyma, right lobe

stomach containing contrast medium

inferior vena cava

aorta

spleen

vertebral body

lower lobe, lung

lower lobe of lung

posterior segments of rib

spinous process

spinal canal

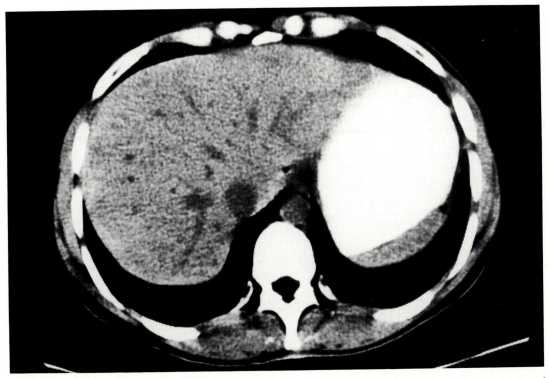

5-38 Abdominal CT cross section 1 of upper abdomen, showing liver, stomach, and superior tip of spleen. Barium sulfate in the stomach.

5/Upper Abdomen and Liver, Computed Tomography

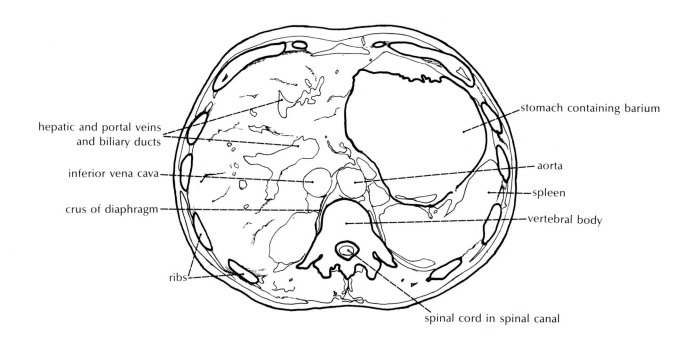

hepatic and portal veins
and biliary ducts

inferior vena cava

crus of diaphragm

ribs

stomach containing barium

aorta

spleen

vertebral body

spinal cord in spinal canal

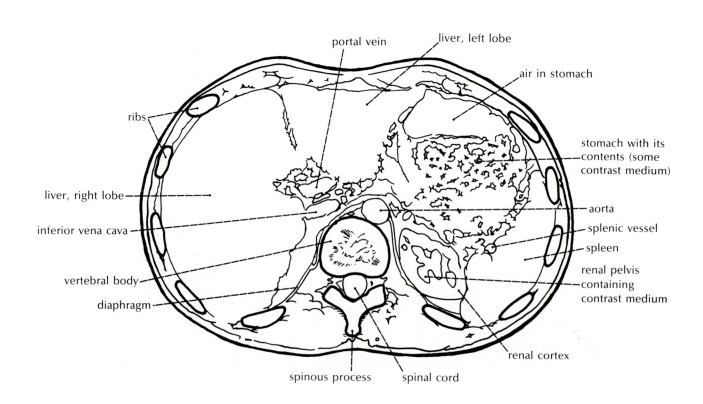

portal vein

liver, left lobe

air in stomach

ribs

stomach with its
contents (some
contrast medium)

liver, right lobe

aorta

inferior vena cava

splenic vessel

spleen

renal pelvis
containing
contrast medium

vertebral body

diaphragm

renal cortex

spinous process

spinal cord

5/Upper Abdomen and Liver, Computed Tomography

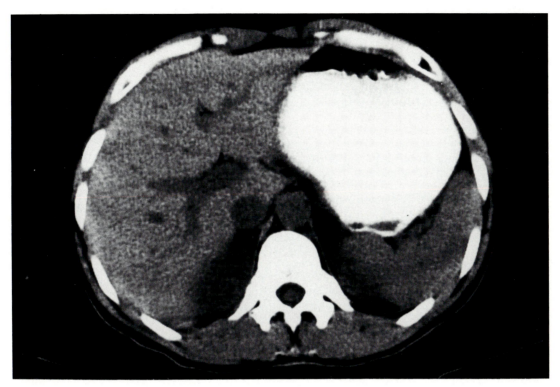

5-39 Abdominal CT cross section 2 of upper abdomen, showing liver, stomach, and spleen. Barium sulfate in the stomach.

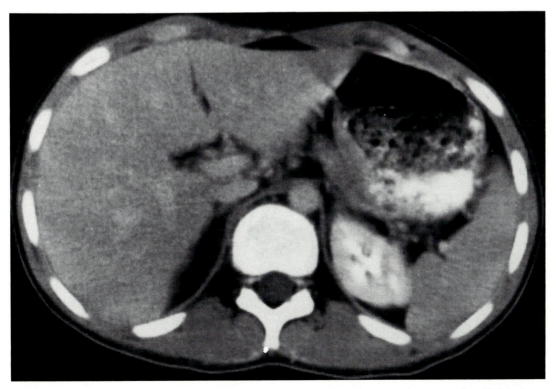

5-40 Abdominal CT cross section 3, showing liver, stomach, spleen, and upper pole of left kidney. Barium sulfate in stomach, water-soluble contrast medium being excreted by the kidneys.

5/Kidneys, Computed Tomography

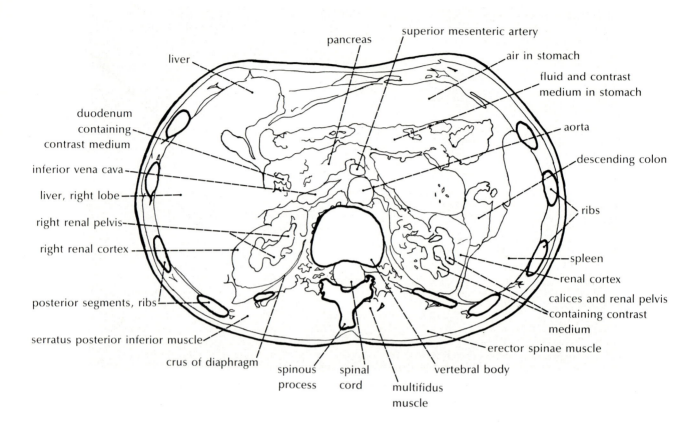

pancreas

superior mesenteric artery

liver

air in stomach

fluid and contrast
medium in stomach

duodenum
containing
contrast medium

aorta

descending colon

inferior vena cava

liver, right lobe

right renal pelvis

right renal cortex

ribs

spleen

renal cortex

calices and renal pelvis
containing contrast
medium

posterior segments, ribs

serratus posterior inferior muscle

erector spinae muscle

crus of diaphragm

spinous
process

spinal
cord

vertebral body

multifidus
muscle

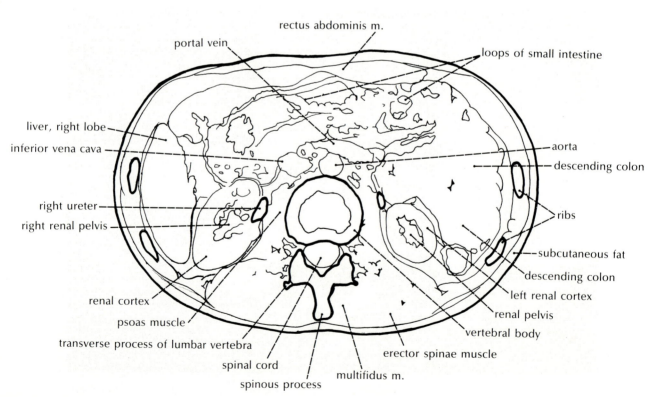

rectus abdominis m.

portal vein

loops of small intestine

liver, right lobe

inferior vena cava

aorta

descending colon

ribs

right ureter

right renal pelvis

subcutaneous fat

descending colon

left renal cortex

renal pelvis

renal cortex

vertebral body

psoas muscle

erector spinae muscle

transverse process of lumbar vertebra

spinal cord

multifidus m.

spinous process

5/Kidneys, Computed Tomography

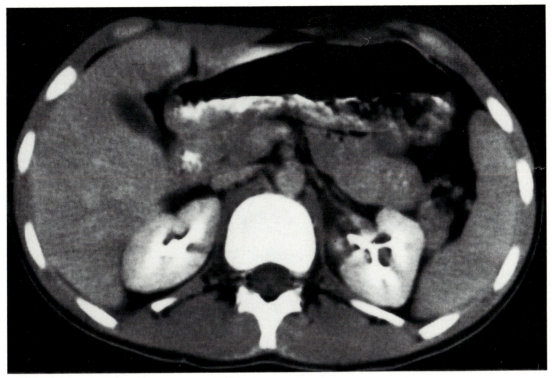

5-41 Abdominal CT cross section 4, showing both kidneys and lower portions of liver and spleen.

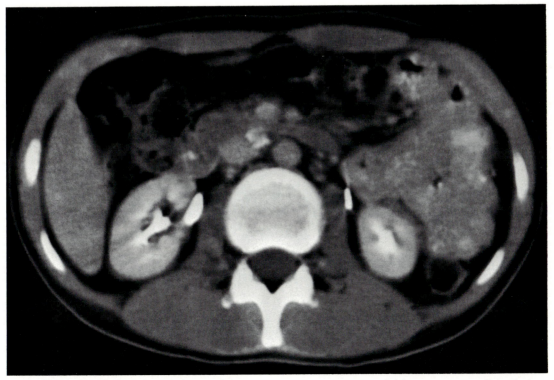

5-42 Abdominal CT cross section 5, inferior to section 4. Water-soluble contrast medium being excreted by the kidneys.

5/Mid Abdomen, Computed Tomography

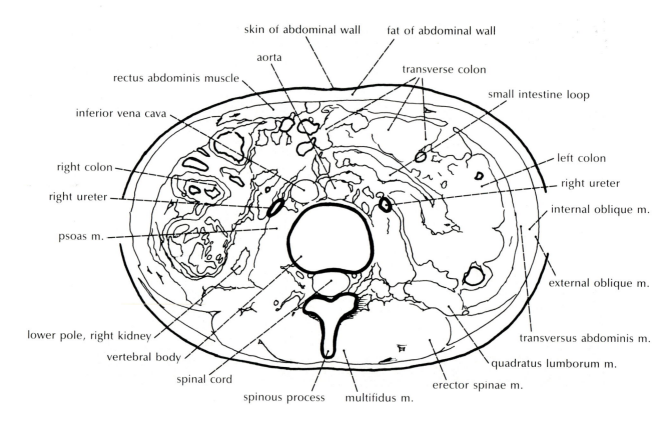

skin of abdominal wall · fat of abdominal wall · aorta · transverse colon · rectus abdominis muscle · small intestine loop · inferior vena cava · right colon · left colon · right ureter · right ureter · internal oblique m. · psoas m. · external oblique m. · lower pole, right kidney · transversus abdominis m. · vertebral body · quadratus lumborum m. · spinal cord · erector spinae m. · spinous process · multifidus m.

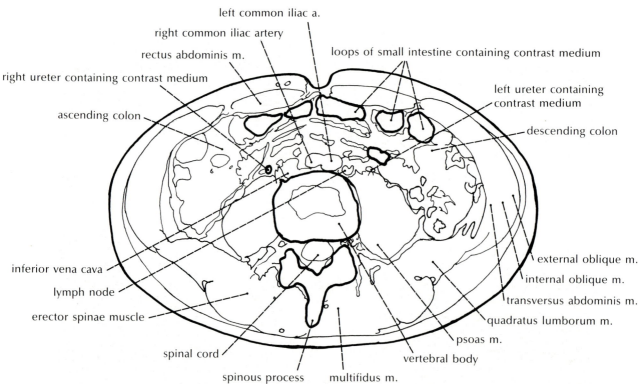

left common iliac a. · right common iliac artery · rectus abdominis m. · loops of small intestine containing contrast medium · right ureter containing contrast medium · left ureter containing contrast medium · ascending colon · descending colon · inferior vena cava · external oblique m. · lymph node · internal oblique m. · erector spinae muscle · transversus abdominis m. · quadratus lumborum m. · spinal cord · psoas m. · spinous process · vertebral body · multifidus m.

5/Mid Abdomen, Computed Tomography

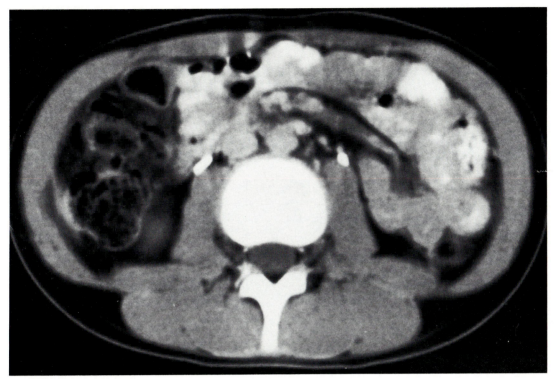

5-43 Abdominal CT cross section 6, mid abdomen below level of kidneys. Barium sulfate in intestinal loops, water-soluble contrast medium in the ureters.

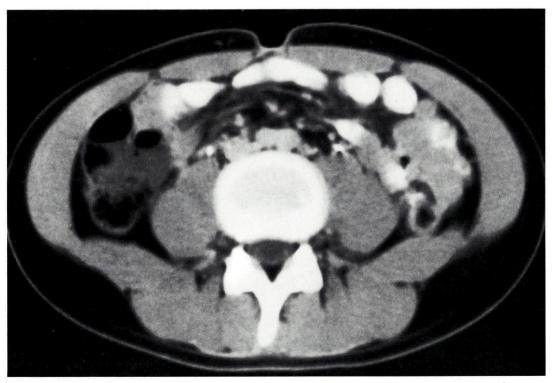

5-44 Abdominal CT cross section 7, mid abdomen, inferior to section 6. Contrast media as in section 6.

5/Pelvic Brim and Lower Pelvis, Male, CT

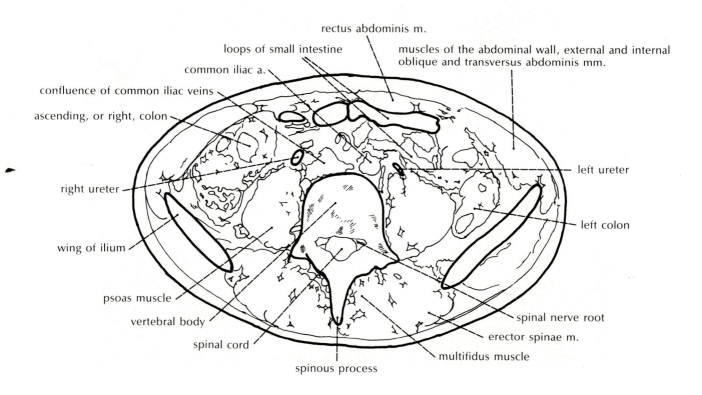

rectus abdominis m.

loops of small intestine

common iliac a.

muscles of the abdominal wall, external and internal oblique and transversus abdominis mm.

confluence of common iliac veins

ascending, or right, colon

left ureter

right ureter

left colon

wing of ilium

psoas muscle

vertebral body

spinal cord

spinous process

spinal nerve root

erector spinae m.

multifidus muscle

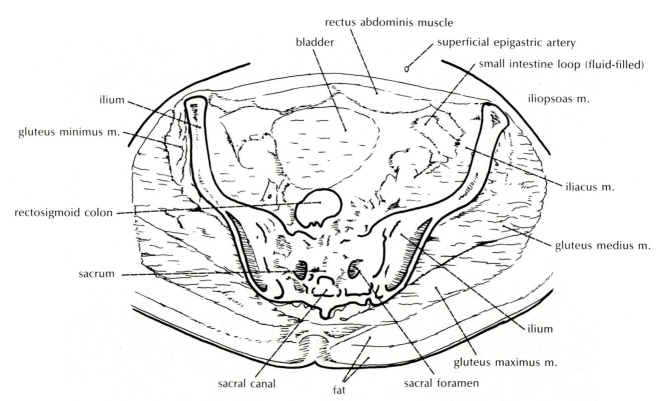

rectus abdominis muscle

bladder

superficial epigastric artery

small intestine loop (fluid-filled)

iliopsoas m.

ilium

gluteus minimus m.

iliacus m.

rectosigmoid colon

gluteus medius m.

sacrum

ilium

gluteus maximus m.

sacral canal

fat

sacral foramen

5/Pelvic Brim and Lower Pelvis, Male, CT

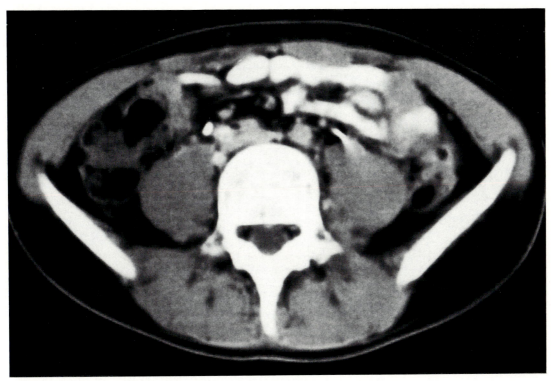

5-45 Abdominal CT cross section 8, just below the pelvic brim. Barium sulfate in intestinal loops, water-soluble contrast medium in the ureters.

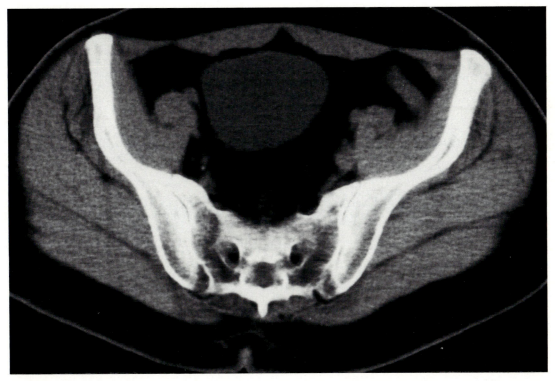

5-46 Abdominal CT cross section 9 in the male pelvis. Urine without contrast medium in the bladder.

5/Lower Pelvis, Male, CT

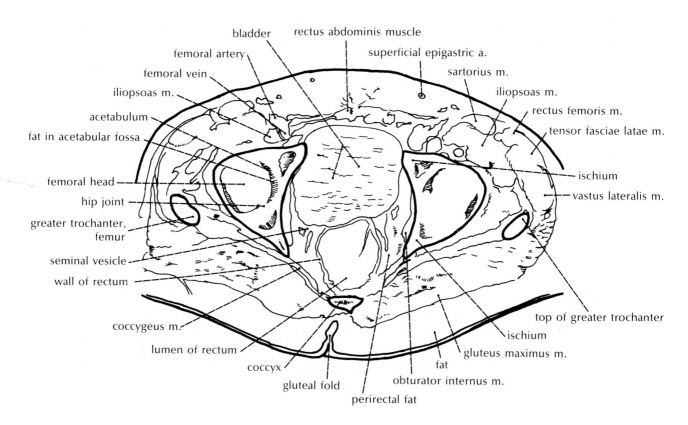

bladder
rectus abdominis muscle
femoral artery
superficial epigastric a.
femoral vein
sartorius m.
iliopsoas m.
iliopsoas m.
acetabulum
rectus femoris m.
fat in acetabular fossa
tensor fasciae latae m.
ischium
femoral head
vastus lateralis m.
hip joint
greater trochanter, femur
seminal vesicle
wall of rectum
top of greater trochanter
coccygeus m.
ischium
lumen of rectum
gluteus maximus m.
coccyx
fat
gluteal fold
obturator internus m.
perirectal fat

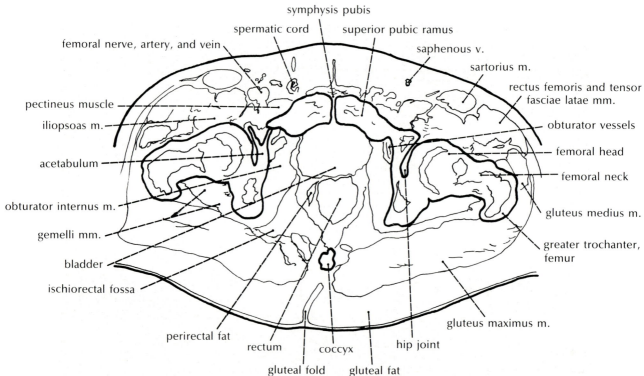

symphysis pubis
spermatic cord
superior pubic ramus
femoral nerve, artery, and vein
saphenous v.
sartorius m.
rectus femoris and tensor fasciae latae mm.
pectineus muscle
obturator vessels
iliopsoas m.
femoral head
acetabulum
femoral neck
obturator internus m.
gluteus medius m.
gemelli mm.
greater trochanter, femur
bladder
ischiorectal fossa
gluteus maximus m.
perirectal fat
rectum
coccyx
hip joint
gluteal fold
gluteal fat

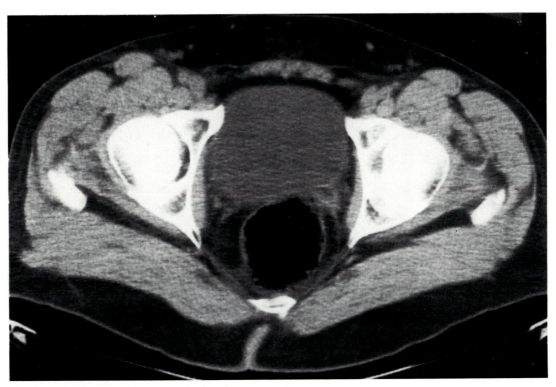

5-47 Abdominal CT cross section 10, lower male pelvic level showing urine-filled bladder (without contrast medium) and rectum. Acetabular level of pelvis.

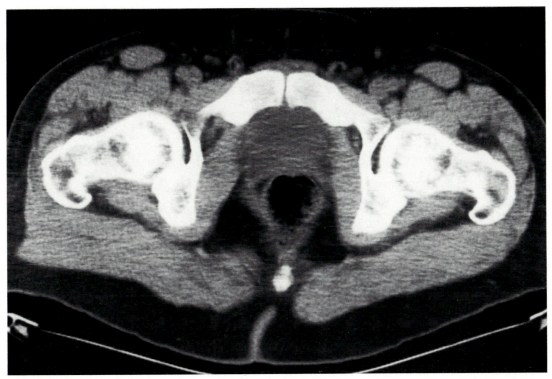

5-48 Abdominal CT cross section 12, lower male pelvis at level of femoral head and neck and symphysis pubis.

5/Lower Pelvis, Female, CT

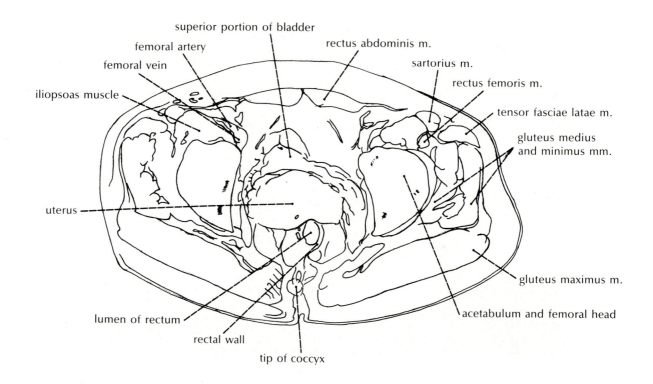

superior portion of bladder
femoral artery
femoral vein
iliopsoas muscle
rectus abdominis m.
sartorius m.
rectus femoris m.
tensor fasciae latae m.
gluteus medius and minimus mm.
uterus
gluteus maximus m.
acetabulum and femoral head
lumen of rectum
rectal wall
tip of coccyx

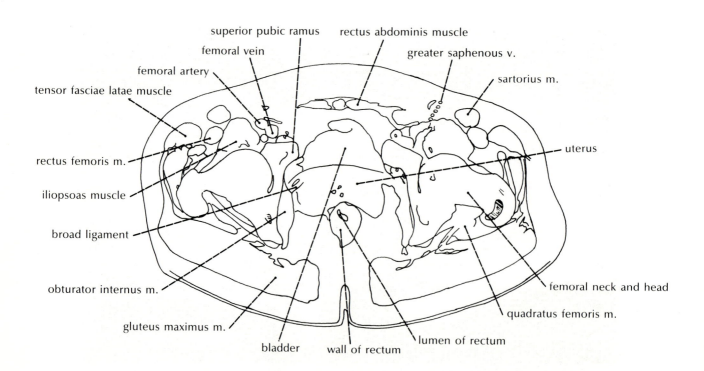

superior pubic ramus
femoral vein
femoral artery
tensor fasciae latae muscle
rectus abdominis muscle
greater saphenous v.
sartorius m.
rectus femoris m.
uterus
iliopsoas muscle
broad ligament
obturator internus m.
femoral neck and head
gluteus maximus m.
quadratus femoris m.
bladder
wall of rectum
lumen of rectum

5/Lower Pelvis, Female, CT

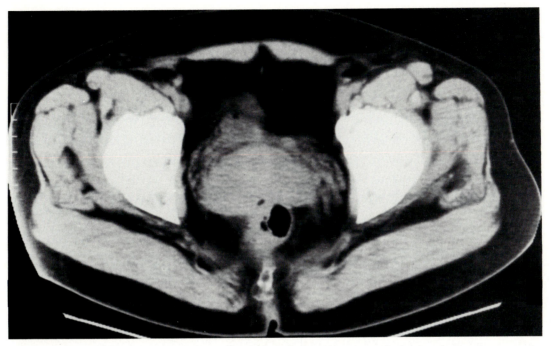

5-49 Abdominal CT cross section 11, lower female pelvis at acetabular level showing superior portion of bladder, uterus, and rectum. No contrast media.

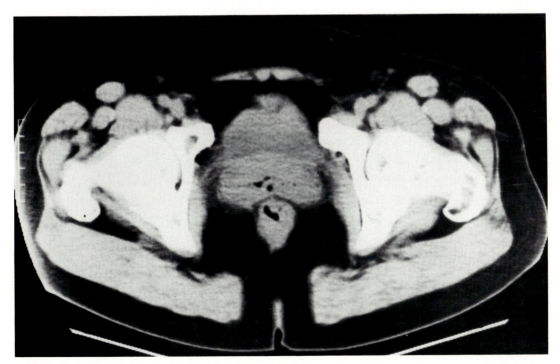

5-50 Abdominal CT cross section 13, at level of femoral head and neck showing bladder, uterus, and rectum. No contrast media.

5/Lowermost Pelvis, Male, CT

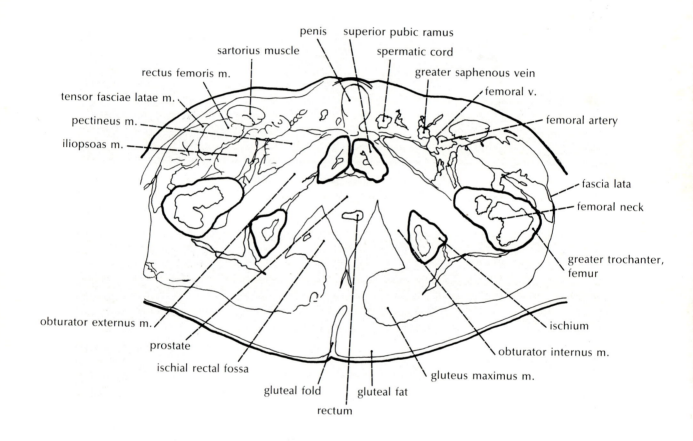

penis superior pubic ramus

sartorius muscle spermatic cord

rectus femoris m. greater saphenous vein

tensor fasciae latae m. femoral v.

pectineus m. femoral artery

iliopsoas m.

fascia lata

femoral neck

greater trochanter, femur

obturator externus m. ischium

prostate obturator internus m.

ischial rectal fossa gluteus maximus m.

gluteal fold gluteal fat

rectum

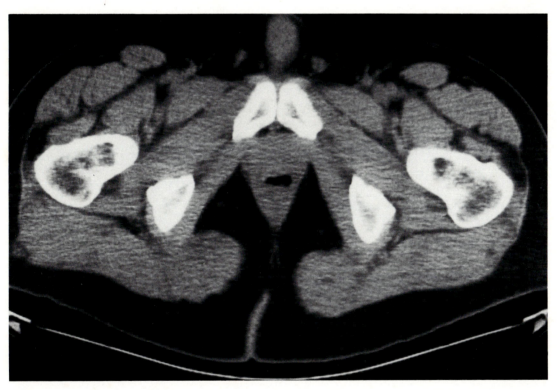

5-51 Abdominal CT cross section 14, lowermost male pelvis, level of femoral intertrochanteric region.